International Congress on Theoretical and Applied Mathematics, Physics and Chemistry 2016

AA006541

Journal of Physics: Conference Series Volume 725

Bandung, Indonesia
23 – 24 April 2016

Editor:

Ford Lumban Gaol

ISBN: 978-1-5108-2676-2
ISSN: 1742-6588

Printed from e-media with permission by:

Curran Associates, Inc.
57 Morehouse Lane
Red Hook, NY 12571

Some format issues inherent in the e-media version may also appear in this print version.

Copyright© (2016) by the Institute of Physics
All rights reserved. The material featured in this book is subject to
IOP copyright protection, unless otherwise indicated.

Printed by Curran Associates, Inc. (2016)

For permission requests, please contact the Institute of Physics
at the address below.

Institute of Physics
Dirac House, Temple Back
Bristol BS1 6BE UK

Phone: 44 1 17 929 7481
Fax: 44 1 17 920 0979

techtracking@iop.org

Additional copies of this publication are available from:

Curran Associates, Inc.
57 Morehouse Lane
Red Hook, NY 12571 USA
Phone: 845-758-0400
Fax: 845-758-2633
Email: curran@proceedings.com
Web: www.proceedings.com

TABLE OF CONTENTS

Journal of Physics: Conference Series **725** (2016) 011001 doi:10.1088/1742-6596/725/1/011001

Preface

It's a great pleasure to present the proceedings of the 2016 International Congress on Theoretical and Applied Mathematics, Physics and Chemistry (The Science 2016), which was held at the The Panghegar Hotel - Bandung, Indonesia, during 23 - 24 April 2016.

This year, we received 37 submission and after the review and editorial revision process we accepted 11.

The theme of Science 2016 is *The Beauty in Science: From Concept to Technology.*

The conference aims to bring together scholars, leading researchers and experts from diverse backgrounds and applications in mathematics, physics and chemistry. Special emphasis is placed on promoting interaction between the experimental and applied communities, so that a high level of exchange in new and emerging areas within all areas is achieved.

We would like to express our sincere gratitude to all in the Technical Program Committee who have reviewed the selected papers and given technical and/or editorial comments for those papers

We do hope that this special issue will give positive impact to the mathematics, physics and chemistry research community.

The Guest Editor of the The Science 2016

Ford Lumban Gaol
Bina Nusantara University

Content from this work may be used under the terms of the Creative Commons Attribution 3.0 licence. Any further distribution of this work must maintain attribution to the author(s) and the title of the work, journal citation and DOI.
Published under licence by IOP Publishing Ltd

Journal of Physics: Conference Series **725** (2016) 011001 doi:10.1088/1742-6596/725/1/011001

Guest Editor Biographies

Ford Lumban Gaol received the B.Sc. in Mathematics, Master of Computer Science. and the Doctor in Computer Science from the University of Indonesia, Indonesia in 1997, 2001 and 2009, respectively.

He is currently Associate Professor Informatics Engineering and Information System, Bina Nusantara University www.binus.ac.id. He is the Chair of PhD Program and Research Interest Group Leader "Advance System in Computational Intelligence & Knowledge Engineering " (IntelSys) Bina Nusantara University.
He was the co-founder of JIBES School of Computer and Informatics Management www.jibes.ac.id.

Dr Ford is the officer of IEEE Indonesia section for International and Professional Activities http://ieee.web.id/indonesia/officers/. He is the Chair SERSC: Science & Engineering Research Support soCiety Indonesia Section. http://www.sersc.org/organization.php Dr Ford was the ACM Indonesia Chapter Chair
http://campus.acm.org/public/chapters/geo_listing/index.cfm?rabbr=Indonesia&inus=0&ct=Professional

Dr Ford involved with some project relate with Technology Alignment in some of multinational company like Astra, United Tractors, Telkom, Sony Erickson.

For International collaboration, Dr Ford is the recipient of IEEE Visiting Professor to Hong Kong University in 2011. For 2012, Dr Ford already received confirmation form IAMSIE Hong Kong to collaborate with R&D with some Manufacturing Companies in HongKong.

Journal of Physics: Conference Series **725** (2016) 011001 doi:10.1088/1742-6596/725/1/011001

Keynote Speakers

1. Prof Tokuro Matsuo, Ph.D
Professor, Advanced Institute of Industrial Technology, Japan
Guest Professor, Bina Nusantara University, Indonesia
Project Professor, Nagoya Institute of Technology, Japan

Title: Computational Method in Material Design

Artificial Intelligence(AI)-based search methods provide a lot of possibility to solve complex problem. However, except for logical reasoning, it cannot be used to analyze the detail of models because AI-based methods provide just roughly result through simulations. Traditionally, operation research(OR)-based methods may have a strong power to analyze more precise and concrete detail of local effects of systems. However, in a real world, a system sometimes has an uncertain temper like a cat. In material science and engineering, manufacturing of semiconductor is speeded up year by year because of competition with each implementing company. Thus, a lot of material scientists are annoyed with such actual situation to design new material of semiconductor. In this talk, I introduce our research contribution to the material design on adhesive design to find appropriate compositions. We employ multiple mixed-techniques including linear programming, nonlinear programming, reasoning methods, qualitative analysis and interpolation algorithm. Also, I show our developed software that is used in factories. It is used to find appropriate types of material, quantity of each material, and their combinations.

Speaker Bio

Tokuro Matsuo is a full professor at Advanced Institute of Industrial Technology since 2012. He received the doctor degree of engineering from Dept. of Computer Science at Nagoya Institute of Technology in 2006. He is a guest professor at Bina Nusantara University, Indonesia since 2015; a research project professor of Collective Intelligence Research Center at Nagoya Institute of Technology, Japan since 2015; a research fellow of SEITI in Central Michigan University, USA since 2010; and an executive director of International Institute of Applied Informatics since 2010. He was a visiting researcher at University of California at Irvine in 2010-2011; was a research fellow at Shanghai University between 2010 to 2013; and was a research project professor of Green Computing Research Center at Nagoya Institute of Technology

Journal of Physics: Conference Series **725** (2016) 011001 doi:10.1088/1742-6596/725/1/011001

between 2011 to 2014. His current research interests include electronic commerce and business, service science and marketing, business management, artificial intelligence, material informatics, tourism informatics, convention research, and incentive design on e-services. Some of his researches are presented in the top international conferences on AAAI, IEEE CEC, AAMAS, and WWW.

He chaired a lot of international conferences including IEEE/ACIS SNPD 2009, 2012 and 2014, IEEE/ACIS ICIS 2010 and 2013, IEEE IWEA 2007-2012, ACAN 2005-2012, and AAI 2012-2016. He gave over 70 keynotes and invited talk at international conferences, symposia, and seminars. He also received over 40 awards and research grants from research foundations, company and Government. He is also commissioned as Japan Conference Ambassador, Kumamoto City MICE Ambassador, and Adviser of Information Promotion of Japan.

2. Prof. Yiyu Cai, Ph.D
Nanyang Technological University (NTU).

Title : Virtual Reality: Research, Development and Applications.

In March 2014, Facebook announced the USD$2 Billion acquisition of the Oculus Rift VR. In Jan 2015, Microsoft revealed its Augment Reality product HoloLens. 2016 is considered The Year of Virtual Reality (VR). VR, however, is not new. Back to the mid of 1960s, Ivan Sutherland, the inventor of the first VR Head Mounted Display (HMD), described "The screen is a window through which one sees a virtual world. The challenge is to make that world look real, act real, sound real, feel real". In this talk, the keynote speaker will share the research and development of Virtual Reality at Nanyang Technological University, Singapore. In particular, fidelity modelling, real-time interaction, immersive visualization, and natural user interface will be detailed. This is followed by the presentation of VR applications in medicine, education, and engineering. Future direction of VR research and development will be discussed.

Speaker Bio

Professor Yiyu Cai directs the Computer-aided Engineering Labs, and heads the Strategic Program of Virtual-Reality and Softcomputing with The School of Mechanical & Aerospace Engineering, Nanyang Technological University (NTU). He

Journal of Physics: Conference Series **725** (2016) 011001 doi:10.1088/1742-6596/725/1/011001

has over 20 years' experience in VR research and applications development. His research has been supported by both government and public funding agencies. He has been keynote speaker and invited speaker for numerous international and regional conferences. He was the former Co-president of The International Simulation and Gaming Association. He teaches VR, Computer Graphics and Mechatronics in NTU.

3. Zhao Shengdong, Ph.D
National University of Singapore

Title : Innovation Through Principles

Human-computer Interaction (HCI) is a very broad field without a single clear focus - there are a great number of HCI researchers from multiple disciplines working on a wide variety of problems that seem to have little connection with each other. However, a careful examination of these multiple approaches to problem solving seems to indicate that there are a number of distinctive, underlying design principles employed by researchers. In this talk, I try to describe three commonly seen design principles in HCI innovations, including "Enhance Directness, Facilitate Transition, Balance Humans and Computers", and unpack these principles using projects from the NUS-HCI lab.

Speaker Bio

Dr. Shengdong Zhao is an Assistant Professor in the Department of Computer Science of the National University of Singapore where he founded the NUS-HCI research lab in 2009. He earned his Masters and PhD degrees at the University of California at Berkeley and the University of Toronto, respectively. Shengdong has a wealth of experience in developing new interface tools and applications. ACM Interaction Magazine mentioned him as one of the most prolific authors in Asia for the ACM SIGCHI conference in the last six years. Shengdong frequently served on the program committees in top HCI conferences. He was a co-chair of the Interaction Using Specific Capabilities or Modalities subcommittee for SIGCHI in the last two years. Shengdong was also the recipient of the NUS Young Investigator Award.

Journal of Physics: Conference Series **725** (2016) 011001 doi:10.1088/1742-6596/725/1/011001

The 2016 International Conference
on Data, Internet & Education Technologies
(The DIET 2016)
The 2016 International Congress
on Theoretical and Applied Mathematics, Physics & Chemistry
(The Science 2016)
The 3rd International Conferences
Game Technology 2016
(VisioGame 2016)
Royal Panghegar Hotel
Indonesia, 23 - 24 April 2016
www.graphgame.org

Peer review statement

All papers published in this volume of *Journal of Physics: Conference Series* have been peer reviewed through processes administered by the proceedings Editors. Reviews were conducted by expert referees to the professional and scientific standards expected of a proceedings journal published by IOP Publishing.

Content from this work may be used under the terms of the Creative Commons Attribution 3.0 licence. Any further distribution of this work must maintain attribution to the author(s) and the title of the work, journal citation and DOI.
Published under licence by IOP Publishing Ltd

Study synthesis of $(La_{1-x}Gd_x)Ba_2Cu_3O_{7-\delta}$ superconductors at low temperature

M Sumadiyasa[1], I G A Putra Adnyana[2], I G A Widagda[3], W G Suharta[4]

[1, 2, 3, 4] Department of Physics, Faculty of Mathematics and Natural Sciences, Udayana University

E-mail: sumadiyasa64@gmail.com

Abstract. This study has been conducted on synthesis of $(La_{1-x}Gd_x)Ba_2Cu_3O_{7-\delta}$ superconductors with the substitution of Gadolinium (Gd) (x = 0.25, 0.5, 0.75, 1) and sintering temperature (700-960°C) variation using a wet-mixing method. Characterization is done by XRD, RAMAN and SEM. XRD characterization results of all samples have shown sharp peaks which indicate that the sample had crystallized well. Search match results showed an impurity phase such as $BaCO_3$, CuO and $BaCuO_2$. Rietveld analyses for all samples gave decreasing lattice parameters (*a*-axis from 3.9069 to 3.8936 Å, *b*-axis from 3.9231 to 3.9055 Å and *c*-axis from 11.8489 to 11.6597 Å) with the addition of Gd contents (from 0.25 to 1.00). Such addition also caused a decrease of lattice parameters (*a*-axis from 4,901 to 3.8975 Å, *b*-axis from 3.9658 to 3.8986 Å and *c*-axis from 11.7254 to 11.6758 Å) with the addition of sintering temperature (from 700 to 960°C). Characterization of FTIR seen the bending vibration absorption by CO_3^{2-}, absorbs mode apical oxygen of La (Gd) -O-Cu(2) and Cu(1)-O(1)-Cu(2). The addition of sintering temperatures also increases the intensity of the superconducting phase, reduce the intensity of an impurity phase (based on search results match) and increase the particle size (based on SEM characterization and Scherrer calculation).

1. Introduction

Researchers interested in conducting research of $GdBa_2Cu_3O_7$ superconductor materials (Gd-123), due to application of the material can penetrated in various aspects of life. Some of these applications are superconducting quantum interference device (SQUID) and manufacture of fast trains by using magnetic levitation.

The discovery of ceramic superconductor on LBCO systems by Bednorz and Muller [1] with a critical temperature of about 30 K has inspired researchers to conduct further research. Wu, et.al. [2] has reported YBCO superconductor with a critical temperature of about 90 K. Then replacing the block Y to block RE has increased the critical temperature to 93 K. The rare earth element (RE) is a rare earth element such as Nd, Eu, Gd, Dy, Sm and others. These elements are magnetic materials with the element RE on the position (½, ½, ½) in Tri-perovskite structure.

In the synthesis process of $GdBa_{2-x}Cu_3O_{7-\partial}$ superconductor, some researchers have reported a variety of methods, like more flux variations, variations of doping, variations of temperature and sintering time. The addition of Zr doping has reported by Caixuan Xu, et. al. [3] on $GdBa_{2-x}Cu_3O_{7-\partial}$ superconductor using the method of melting at a temperature of 1030°C modified by the top-seed and the annealing process using oxygen and argon gas.

Content from this work may be used under the terms of the Creative Commons Attribution 3.0 licence. Any further distribution of this work must maintain attribution to the author(s) and the title of the work, journal citation and DOI.
Published under licence by IOP Publishing Ltd

Additional Zr doping has produced the critical current of 105 A/cm^2 at a temperature of 77 K. E Ban, et.al. provided dope of ZnO and ZrO_2 on $(Nd_{0.33}Eu_{0.38}Gd_{0.28})Ba_2Cu_3O_x$ superconductor to improve the critical current density [4], while T Goto, et.al. synthesized the same material with added 0.1% Zr and the addition of 0.1% and 0.1% oxygen with an argon gas solution spinning method [5]. Giving dope of Ag on $GdBa_{2-x}Cu_3O_{7-\partial}$ superconductor to fill the porosity, grain boundary, the normal state resistivity and weak-link effects have been reported by N Sakai [6]. While the infiltration methods by giving doping Ag was used by N Mori, et. al. for crystal growth $YBa_{2-x}Cu_3O_{7-\partial}$ [7].

S. Nariki, et.al. has conducted research to get a large grain size of $GdBa_{2-x}Cu_3O_{7-\partial}$ superconductor with added $BaCeO_3$, 20 % Ag_2O and 0.5 % Pt and drained 1 % oxygen. That has produced samples with a grain size of 25 to 33 mm [8]. Research to get large particle size is also done by Y Kimura, et. al. [9] with added a mixture of compounds Bi-Sn-Cd.

In this study observed the effect of Gd substitution on LBCO to form $La_{1-x}Gd_xBa_2Cu_3O_{7-\delta}$ superconductors. The synthesis process using wet-mixing method and sintering at low temperatures, to get homogeneity of grains and to know evolve the growth that superconductor. The contribution of this paper is to give knowledge about the effect of Gd element on LBCO superconductors to the crystal structure. Besides, it aims to produce an effective method of generating homogeneous grain size.

2. Experiment

Samples $(La_{1-x}Gd_x)Ba_2Cu_3O_{7-\delta}$ synthesized using a wet-mixing method with variation of gadolinium (Gd) (x = 0.25, 0.5, 0.75, 1) and sintering temperature (700-960°C). The raw material using Gd_2O_3 powder compound (99.9%), La_2O_3 (99.9%), $BaCO_3$ (99.9%) and CuO (99.9%). The raw materials are weighed according to the nominal composition, then each material dissolved with HNO_3 solution accompanied by stirring for 30 minutes to get homogeneous solutions. All solution mixed and stirred with added sintering temperature slowly (with a temperature of less than 100°C) until crusted. Then, the precipitate heated at 100°C for 1 hour [10, 11]. The results calcined at 650°C for 3 hours and sintering at 700-900°C respectively for 1 hour. The samples were characterized using X-ray diffraction (XRD), fourier transform infrared (FTIR) and scanning electron microscopy (SEM).

3. Results and Discussion

The XRD characterization results of the $(La_{1-x}Gd_x)Ba_2Cu_3O_{7-\delta}$ samples with the variation of x (0.25, 0.5, 0.75, 1) and sintering at a temperature of 800°C for 1 an hour are shown in figure 1. While the sample $(La_{0.5}Gd_{0.5})Ba_2Cu_3O_{7-\delta}$ with sintering temperature variations (700, 800, 900, 960°C), respectively for 1an hour are shown in figure 2. In general, the spectrum shows sharp peaks indicating that crystallization had occurred to the good and perfect. But in some peaks still indicated the existence of peaks that overlap or have not been properly separated. This is due to the sintering temperature is still too low compared ideally sintering temperature (970°C), so that the thermal energy required for the formation of a single phase is not enough. Besides, it takes longer to acquire the heating to growth superconductor phase at a higher intensity.

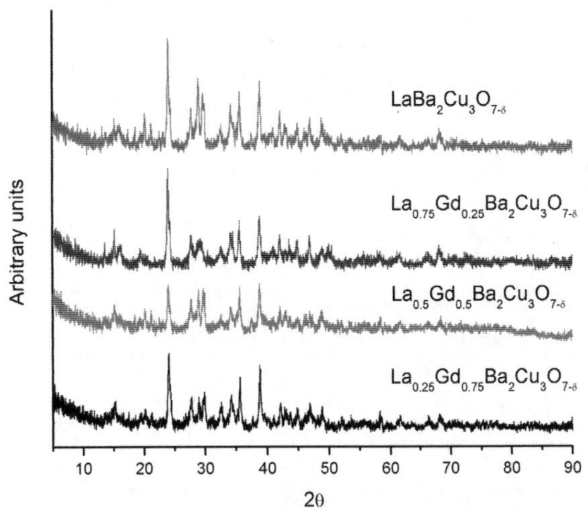

Figure 1. The results of the XRD characterization of $(La_{1-x}Gd_x)Ba_2\,Cu_3O_{7-\delta}$ samples with a variation of x (0.25, 0.5, 0.75, 1), each at 800°C

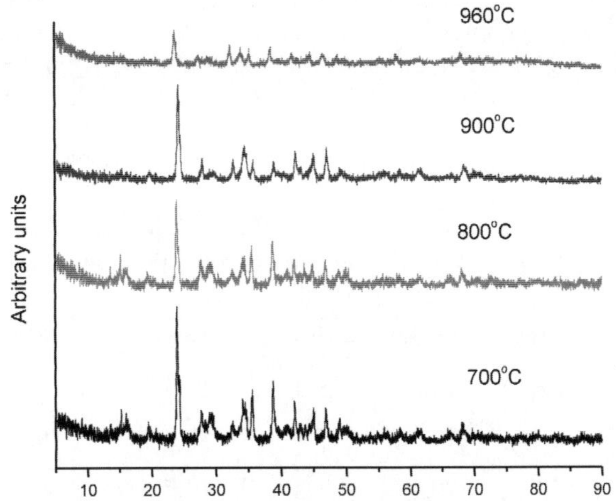

Figure 2. The results of the XRD characterization of $La_{0.5}Gd_{0.5}Ba_2\,Cu_3O_{7-\delta}$ samples with sintering temperature variations (700, 800, 900, 960°C)

The matching of the XRD diffraction peak of the samples has done using Match program. The results of matching were identified impurity phases such as $BaCO_3$, CuO and $BaCuO_2$ especially at angle 2θ of 22 to 35° shows that figure 3 (for x variations) and figure 4 (for sintering temperature variations). Emerge an impurity phase with high intensity, caused by give of sintering temperature is still low, so the energy that is available has not been able to foster a phase $(La_{0.5}Gd_{0.5})Ba_2Cu_3O_{7-\delta}$ omit.

$BaCO_3$ and CuO compounds capable decomposes rapidly, probably caused by wet-mixing synthesis method, in which happen the ionic bond. Decomposition of $BaCO_3$ and CuO impurity followed by the growth of $BaCuO_2$ phase.

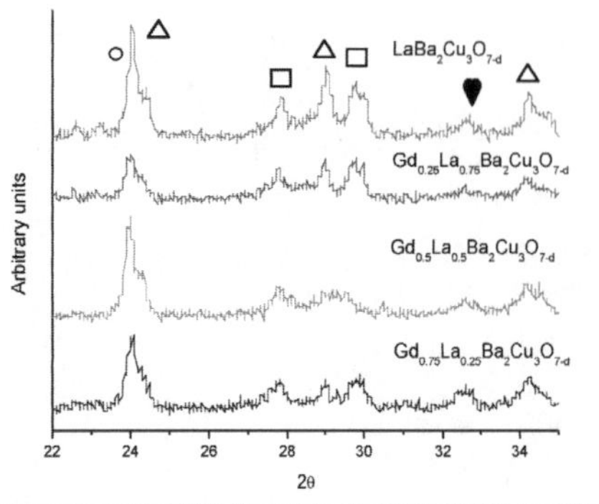

Figure 3. The results of search match of $(La_{1-x}Gd_x)Ba_2Cu_3O_{7-\delta}$ samples with a variation of x (0.25, 0.5, 0.75, 1) from 22 to 35°. Symbol : o ($Gd_{0.5}La_{0.5}Ba_2Cu_3O_{7-\partial}$), Δ ($BaCO_3$), □ (CuO), ♥($BaCuO_2$)

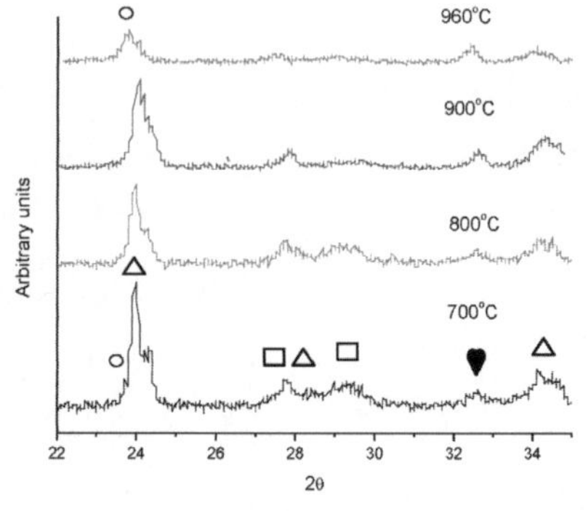

Figure 4. The results of search match of $(La_{1-x}Gd_x)Ba_2Cu_3O_{7-\delta}$ samples with a variation of x (0.25, 0.5, 0.75, 1) from 22 to 35°. Symbol : o ($Gd_{0.5}La_{0.5}Ba_2Cu_3O_{7-\partial}$), Δ ($BaCO_3$), □ (CuO), ♥($BaCuO_2$)

The calculation results of the volume fraction of $(La_{1-x}Gd_x)Ba_2Cu_3O_{7-\delta}$ samples with a variation of x, and $(La_{0.5}Gd_{0.5})Ba_2Cu_3O_{7-\delta}$ samples with variations of sintering temperature were obtained and each showed in table 1 and 2. Volume fraction of $(La_{0.5}Gd_{0.5})Ba_2Cu_3O_{7-\delta}$ and $BaCuO_2$ phases increase with addition of sintering temperature, while $BaCO_3$ and CuO phases as impurities decrease.

Table 1. The volume fraction of $(La_{1-x}Gd_x)Ba_2Cu_3O_{7-\delta}$ samples with variations of x

x	LGBCO (%)	$BaCO_3$ (%)	$BaCuO_2$ (%)	CuO (%)	Other impurity (%)
0.25	55.3	20.3	3.5	14.8	6,1
0.50	56.4	19.6	3.8	15.4	4.8
0.75	56.1	19.4	3.6	15.8	5.1
1.00	55.8	20.1	3,2	14.7	6.2

Table 2. The volume fraction of $(La_{0.5}Gd_{0.5})Ba_2Cu_3O_{7-\delta}$ samples with variations of sintering temperature (700, 800, 900, 960°C)

T (°C)	LGBCO (%)	$BaCO_3$ (%)	$BaCuO_2$ (%)	CuO (%)	Other impurity (%)
700	53.1	21,2	3,1	17,2	5.4

800	56.4	19,6	3.8	15.4	4.8
900	61.4	18.2	4.1	12.1	4.2
960	66.7	16.4	4.5	9.3	3.1

Rietveld refinements were performed using *Rietica* [12] by employing Newton-Raphson strategy-Voight peak shape function. The refinement results showed that the addition of Gd content from 0.25 to 1 resulted in a decrease of the lattice parameter (*a*-axis from 3.9069 to 3.8936 Å, *b*-axis from 3.9231 to 3.9055 Å and *c*-axis from 11.8489 to 11.6597 Å) as shown in figure 5. The decline in values of lattice parameter is caused by the difference of ionic radii Gd and La, where ionic radii of La (117.2 pm) larger than Gd (107.8 pm) [13].

The addition of sintering temperature from 700 to 960°C resulted in a decrease of lattice parameter (*a*-axis from 4.901 to 3.8975 Å, *b*-axis from 3.9658 to 3.8986 Å and *c*-axis from 11.7254 to 11.6758 Å) as shown in figure 6. The decline in values of lattice parameter is caused by loss of oxygen, hydrogen and nitrogen. The reduces of oxygen, hydrogen and nitrogen resulted in the distance between the superconducting blocks closer. This indicates that the superconductivity is better.

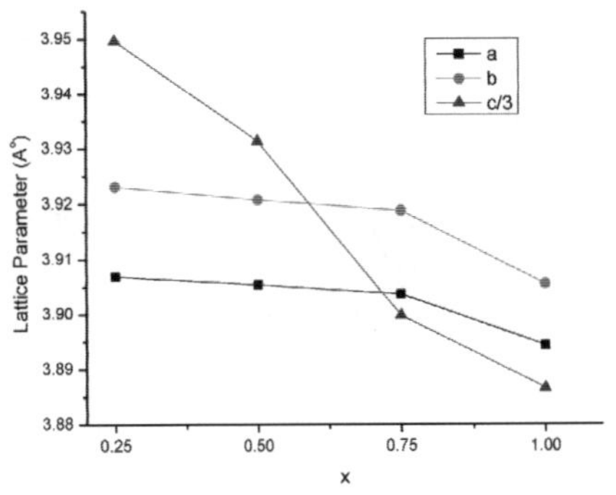

Figure 5. The curve as relationship of the lattice parameter values and x

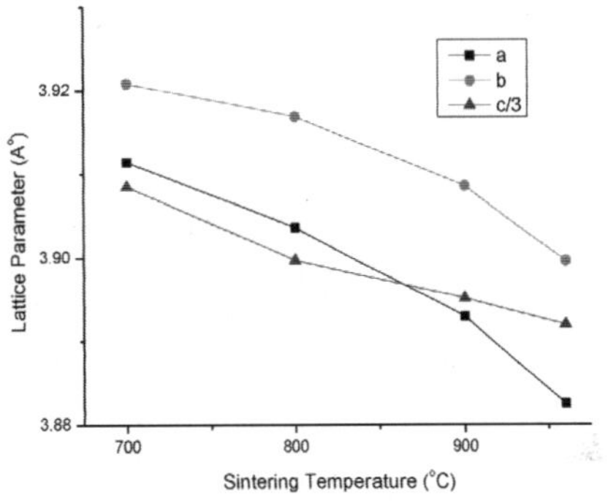

Figure 6. The curve as the relationship of the lattice parameter values and the sintering temperature variations

The results of the FTIR characterization of $(La_{0.5}Gd_{0.5})Ba_2Cu_3O_{7-\delta}$ samples with sintering temperature variations (700, 800, 900°C) were shown in figure 7. The results of search match supported by the results of FTIR measurements.

The absorption peaks at 3374.23 and 1611.97 cm^{-1} attributed to stretching vibration of the O-H bond and the bending vibration of the H-O-H respectively [14, 15], from water molecules on the external surface of the samples during handling to record the spectra. Also, the stretching C=O vibration at about 1750 cm^{-1} confirm form the sample indicating ordinate of the carbonyl group to the metal [16, 17]. The peak shown at about 580 cm^{-1} in the spectrum is attributed to the stretching frequency at 3422 cm^{-1} and a weak asymmetric band at 1627 cm^{-1} support the presence of the OH-group due to the absorption of water by nanoparticle during sample preparation. The presence of two M-O stretching and bending frequencies at about 690 cm^{-1}, 854 cm^{-1} respectively.

Classification of modes associated with displacements of atoms at specific sites in $YBa_2Cu_3O_{7-\partial}$, there is a total of 39 vibrational modes : 15 of which are Raman active with symmetries of 5Ag + 5Big + 4B2g + B3g. The 21 infrared active modes are 7(B1u + B2u + B3u) and there are three acoustic modes. The most interesting result is that the motion of Cu(1) and O(1) do not contribute to Raman active modes, but they are infrared active. The O(4) motion in the "flat" part of the Cu(1) ribbons may contribute to a Raman active mode however [18].

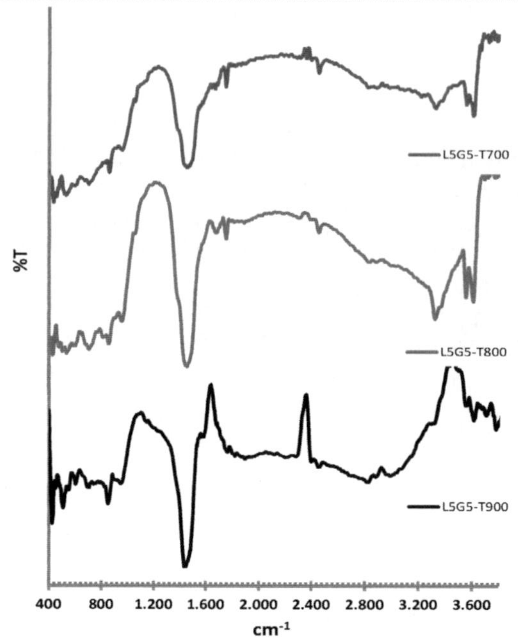

Figure 7. The results of FTIR characterization sample of $(La_{0.5}Gd_{0.5})Ba_2Cu_3O_{7-\delta}$ with sintering temperature variations (700, 800, 900°C)

The Raman characterization results showed in figure 8 (for x variation) and figure 9 (for sintering temperature variation). Six Raman peaks observed : at 210, 248, 338, 451, 505 and 610 cm^{-1} appearing. These lines correspond to A_g modes [19].

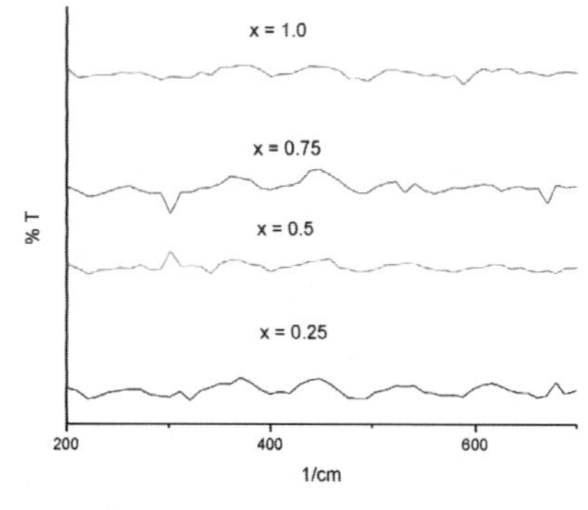

Figure 8. Results of Raman characterization of $(Gd_{1-x}La_x)Ba_2Cu_3O_{7-\delta}$ samples with a variation of x (0.25, 0.5, 0.75, 1.0)

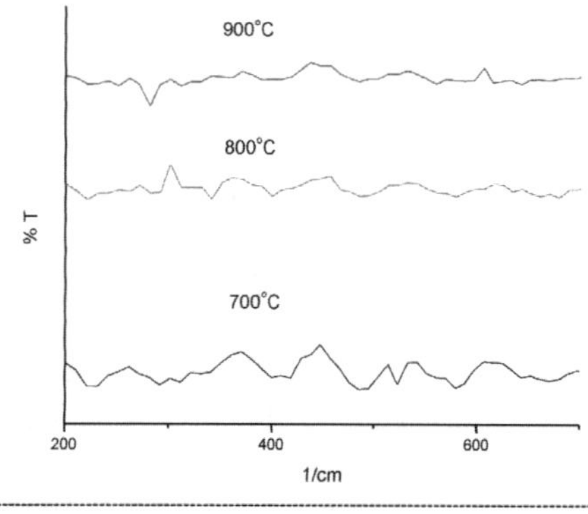

Figure 9. Results of Raman characterization of $Gd_{0.5}La_{0.5}Ba_2Cu_3O_{7-\partial}$ samples with sintering temperature variations (700, 800, 900, 960°C)

 doi:10.1088/1742-6596/725/1/012001

The morphological characterization results of $Gd_{0.5}La_{0.5}Ba_2Cu_3O_{7-\partial}$ samples by using SEM with sintering temperature variations (700, 800, 900°C) showed in figure 10. The sample of $Gd_{0.5}La_{0.5}Ba_2Cu_3O_{7-\partial}$ with sintering temperature of 700°C (figure 10a) shows homogeneous particle size and distributed in the sample with the form of rods, which describes the characteristic of the orthorhombic-shaped, with the lattice parameter values in the direction of the *a*-axis and *b*-axis are almost the same, while the value of the c-axis is almost three times the value of *a* and *b*. Additional sintering temperature of 700, 800 and 900°C resulted in agglomerations, so that granules with a grain of another one already began to coalesce to form a molecular bond (figure 10b and 10c). The calculation result using Scherrer equation obtained the particle size values of 168, 215 and 275 nm respectively at the sintering temperature of 700, 800 and 900°C.

Figure 10a. SEM images of $Gd_{0.5}La_{0.5}Ba_2Cu_3O_{7-\partial}$ samples at 700°C

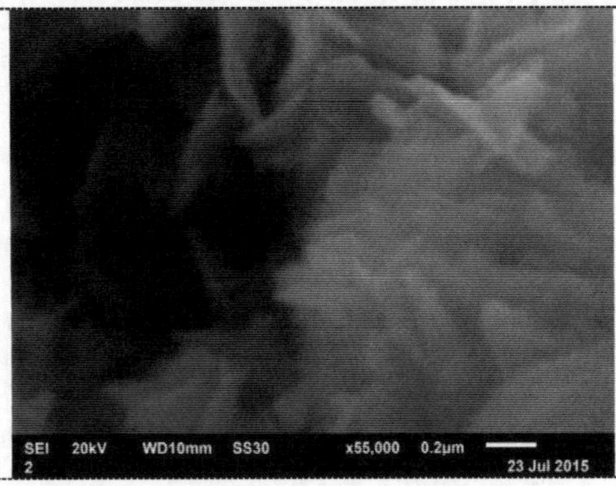

Figure 10b. SEM images of $Gd_{0.5}La_{0.5}Ba_2Cu_3O_{7-\partial}$ samples at 800°C

Figure 10c. SEM images of $Gd_{0.5}La_{0.5}Ba_2Cu_3O_{7-\partial}$ samples at 900°C

4. Conclusion

The synthesis of $(La_{1-x}Gd_x)Ba_2Cu_3O_{7-\delta}$ (for variation of *x* and sintering temperature) samples by using wet-mixing methods has been successfully performed with the highest volume fraction is 66.7%. Additional Gd content resulted decrease of the lattice parameter. Additional sintering temperature resulted decrease of lattice parameter, increasing the volume fraction of superconductor phase, decreasing volume fraction of impurity phase and increasing the particle size. The synthesis using a wet-mixing method is quite effective and efficient to get homogeneous grain size.

Journal of Physics: Conference Series **725** (2016) 012001 doi:10.1088/1742-6596/725/1/012001

Reference

[1] Bednorz J G, Muller K A 1987 *Phys. B* **64** 189
[2] Wu M K, Ashburn J R, Torny C J, Hor P H, Meng R L, Gao L, Huang Z J, Wang Y Q and Chu C W 1987 *Phys. Rev. Lett.* **58** 908
[3] Xu C, A H, Ichihara M, Sakai N, Hirabayashi I, Izumi M 2007 *Physica C* **460-462** 1341-1342
[4] Ban E, Y I, Matsuoka Y, Nishijima G, Watanabe K 2007 *Physica C : Superconductivity* **463-465**: 554-558
[5] Goto T, Sato E, Watanabe K, Nishijima G, Matsui Y, Nagai T, Tsuruta C 2005 *Physica C* **425** 166-170
[6] Sakai N, Inoue K, Nariki S, Hu A, Murakami M, Hirabayashi I 2005 *Physica C* **426-431** 515-519
[7] Mori N, Dateki K, Hirao T, Ogi K 2006 *Physica C* **445-448** 308-311
[8] Nariki S, Sakai N, Murakami M, Hirabayashi I 2006 *Physica C* **445-448** 291-294
[9] Kimura Y, Matsumoto H, Fukai H, Sakai N, Hirabayashi I, Izumi M, Murakami M 2006 *Physica C* **445-448** 408-411
[10] Suharta W G, Mugirahardjo H, Pratapa S, Darminto D, Suasmoro S 2013 *J Supercond. Nov. Magn.* **26** 3209–3214
[11] Suharta W G, Suasmoro S, Pratapa S, Darminto D, *AIP Conference Proceedings* **1555** 62-66
[12] Hunter B A 1997 Rietica for Windows versi 1.7.7
[13] Shannon R D 1976 *Acta Crystallogr A* **32** 751–767
[14] Stuart B H 2004 Infrared Spectroscopy: Fundamentals and Applications, John-Wiey & Sons
[15] Sharma P and Virk H S 2009 *The Open Surface Science Journal* **1** 34-39
[16] Roy S C et. al. 2004 *Ceramics International* http://www.eprint.iitd.ac.in/bitstream/2074 /1447/1/royeff2004.pdf
[17] Ischenko V, Jorg Woltersdorf, Eckhard Pippel, Roberto Koferstein, Hans-Peter Abicht 2007 *Solid State Sciences* **9** 303-309
[18] Macfarlane R M, Hal Rosen and Seki H 1987 *Solid State Communications* **63** 831-834
[19] Hadjiev G, Abrashev M V, Iliev M N and Bozukov L N 1990 *Physica C* **171** 257-264

Journal of Physics: Conference Series **725** (2016) 012002 doi:10.1088/1742-6596/725/1/012002

Approximated optimum condition of second order response surface model with correlated observations

Wayan Somayasa

The Department of Mathematics of Halu Oleo University, Kendari 93232,Indonesia

E-mail: wayan.somayasa@uho.ac.id

Abstract. In the present paper we establish an inference procedure for the eigenvalues of the model matrix of the second-order response surface model (RSM). In contrast to the classical treatment where the sample are assumed to be independently distributed, in this work we do not need such distributional simplification. The confidence region for the unknown vector of the eigenvalues is derived by means of delta method. The finite sample behavior of the convergence result is discussed by Monte Carlo Simulation. We get the approximated distribution of the pivotal quantity of the population eigenvalues as a chi-square distribution model. Next we attempt to apply the method to a real data provided by a mining industry. The data represents the percentage of cobalt (Co) observed over the exploration region.

1. Introduction

The problem of determining the optimum condition of a reproduction process is a common objective in the subject of response surface modelling. Optimum condition means a situation on the experimental region under which the reproduction process attains maximum or minimum yield, cf. Bisgaard and Ankenman [3], Carter, Chincilli and Campbell [5], Myers and Kuhri [10], Myers and Montgomery [11, 12] and Somayasa [17]. In the case of the second-order RSM such an optimum condition is represented by the eigenvalues of the so-called model matrix of the response variable. It can be inferred from the magnitude of the eigenvalues of the model matrix whether or not the process in a fixed stationary point achieves a maximum or a minimum value. This subject has found much attention in the literatures, see e.g. [3, 5, 10, 11, 12, 17].

We study throughout this paper a second-order response surface model on a p-dimensional compact space $\mathbf{I}^p := \Pi_{i=1}^p[a_i, b_i]$, $a_i \leq b_i$ with correlated observations defined as follows:

$$Y_i = \beta_0 + \sum_{j=1}^{p} \beta_j x_{ij} + \sum_{j=1}^{p} \sum_{k=j}^{p} \beta_{jk} x_{ij} x_{ik} + \varepsilon_i, \; i = 1, \ldots, n, \tag{1}$$

where Y_i is the response variable (dependent variable), $\mathbf{x}_i := (x_{i1}, \ldots, x_{ip})^\top \in \Pi_{i=1}^p[a_i, b_i]$ is the experimental condition (independent variable), $\beta := (\beta_0, \beta_1, \ldots, \beta_p, \beta_{11}, \beta_{12}, \ldots, \beta_{pp})^\top \in \mathcal{R}^{(p+1)(p+2)/2}$ is the vector of unknown parameters, and ε_i is the stochastically dependent random error having mean zero and finite positive variance σ_{ii}. Let us introduce a vector

Content from this work may be used under the terms of the Creative Commons Attribution 3.0 licence. Any further distribution of this work must maintain attribution to the author(s) and the title of the work, journal citation and DOI.
Published under licence by IOP Publishing Ltd

$\mathbf{b} := (\beta_1, \ldots, \beta_p)^\top \in \mathcal{R}^p$, and a $p \times p$ symmetric matrix $\mathbf{B}$ defined by

$$\mathbf{B} := \begin{pmatrix} \beta_{11} & 1/2\beta_{12} & \cdots & 1/2\beta_{1p} \\ 1/2\beta_{12} & \beta_{22} & \cdots & 1/2\beta_{2p} \\ \vdots & \vdots & \ddots & \vdots \\ 1/2\beta_{1p} & 1/2\beta_{2p} & \cdots & \beta_{pp} \end{pmatrix} \in \mathcal{R}^{p\times p}.$$

Then Model 1 can be written in vector and matrix form which is more compact and useful for our interest, i.e.

$$Y_i = \beta_0 + \mathbf{x}^\top \mathbf{b} + \mathbf{x}^\top \mathbf{B}\mathbf{x} + \varepsilon_i,\ i = 1, \ldots, n. \tag{2}$$

Hence by using differential technique, the unknown stationary point of the model is given by $\mathbf{x}_0 = -1/2\mathbf{B}^{-1}\mathbf{b}$, see e.g. [8, 10, 11, 12] and the references cited therein. There are many methods and approaches that have been established for investigating the optimality of the point $\mathbf{x}_0$. Among other are ridge and steepest ascent methods studied in the references mentioned above. The application of these methods seem to be limited in that the optimum point is subjected by considering the points lie in a sphere of radius r. We refer interested reader to [8, 10, 11, 12] for a more complete discussion on these methods.

Response surface model with correlated observations are frequently found in spatial data analysis in which data is regarded as a realization of a spatial process $\{Y(\mathbf{x}) : \mathbf{x} \in \mathbf{I}^p\}$ such that for every $\mathbf{x}_i, \mathbf{x}_j \in \mathbf{I}^p$, $Cov(Y(\mathbf{x}_i), Y(\mathbf{x}_j))$ is represented as a function of $(\mathbf{x}_i - \mathbf{x}_j)$, that is $Cov(Y(\mathbf{x}_i), Y(\mathbf{x}_j)) = \gamma(\mathbf{x}_i - \mathbf{x}_j)$ for some function γ. When the process is assumed to be isotropic, then the covariance function depends only on the distance of their position, see e.g. Cressie [6], pp. 32–68 for further discussion regarding stationary spatial process. Study regarding generalized linear model can be found in Arnold [1] and Stapleton [18].

It was investigated in [3, 5, 8, 11, 12] that upon transforming the model to its canonical form it can be seen that the characteristic of the response can be determined by observing the magnitude of the eigenvalues of $\mathbf{B}$. More precisely, if the eigenvalues of $\mathbf{B}$ are all positive, then the response variable attains minimum value at $\mathbf{x}_0$. On the other hand if the eigenvalues of $\mathbf{B}$ are all negative, the stationary point $\mathbf{x}_0$ represents a point in which the response attains maximum value. In the case where the eigenvalues differ in sign, the stationary point is a saddle point. Thus deriving an inference procedure for the eigenvalues of $\mathbf{B}$ is an essential task in response surface methodology. In this paper we construct the asymptotic confidence region as well as the confidence interval for the eigenvalues of the model matrix by applying the so-called delta method. Note that for our results we do not need to assume any specific distribution for the random errors.

The motivating idea in applying the delta method is that the vector of eigenvalues of $\mathbf{B}$ is a differentiable function of $\mathbf{B}$. If the parameter β is estimated by some unbiased estimator such as the least squares estimator $\hat{\beta}_n$ then we get immediately the estimator of the eigenvalues. By adopting some result from Arnold [2] and Pruscha [14] the limiting distribution of the eigenvalues can be obtained.

The derivation of the asymptotic distribution of the estimator of the eigenvalues of $\widehat{\mathbf{B}}^{(n)}$ using delta method is discussed in Section 2. In Section 3 we extend the study to a second-order RSM with correlated responses. The finite distribution of the eigenvalue is investigated in Section 4 using Monte Carlo simulation. The result is then applied to construct the asymptotic confidence region and the confidence interval for the optimal condition of a data obtained from a mining industry, see Section 5. We close this paper in Section 6 with a discussion about several open problems and topics for future researches.

2. Delta method

The application of delta method in RSM was proposed in the work of [3] and [5] for response surface model with independent observations. Let us consider a triangular array of observation Y_{ni} observed on the experimental conditions $\{\mathbf{x}_i | \ i = 1, \ldots, n\}$, $n \geq 1$ such that for all i and $n \geq 1$, Y_{ni} follows the model

$$Y_{ni} = \beta_0^{(n)} + \sum_{j=1}^{p} \beta_j^{(n)} x_{ij} + \sum_{j=1}^{p}\sum_{k=j}^{p} \beta_{jk}^{(n)} x_{ij}x_{ik} + \varepsilon_{ni}, \ i = 1, \ldots, n, \tag{3}$$

where $\{\varepsilon_{ni} | i = 1, \ldots, n, \ n \geq 1\}$ is a sequence of triangular array of correlated random errors with $\mathbf{E}(\varepsilon_{ni}) = 0$ and $Cov(\varepsilon_{ni}, \varepsilon_{nj}) = \sigma_{ij}$, for $i, j = 1, \ldots, n$. We call this model throughout this paper the asymptotic second-order RSM. Let $f_0, f_1, \ldots, f_p$ be linearly independent as functions in $L_2(\lambda^p, \mathbf{I}^p)$ defined by $f_0(\mathbf{x}) = 1$, $f_1(\mathbf{x}) = x_1, \ldots, f_p(\mathbf{x}) = x_p$, for $\mathbf{x} = (x_1, \ldots, x_p)^\top \in \mathbf{I}^p$, where $L_2(\lambda^p, \mathbf{I}^p)$ is the space of squared integrable function on $\mathbf{I}^p$ with respect to the Lebesgue measure λ^p. Let $\mathbf{X}_n$ be an $n \times (p+1)(p+2)/2$ design matrix defined by

$$\mathbf{X}_n = \begin{pmatrix} f_0(\mathbf{x}_1) & f_1(\mathbf{x}_1) & \cdots & f_p(\mathbf{x}_1) & f_{11}(\mathbf{x}_1) & f_{12}(\mathbf{x}_1) & \cdots & f_{pp}(\mathbf{x}_1) \\ \vdots & \vdots & \vdots & \vdots & \vdots & \vdots & & \\ f_0(\mathbf{x}_i) & f_1(\mathbf{x}_i) & \cdots & f_p(\mathbf{x}_i) & f_{11}(\mathbf{x}_i) & f_{12}(\mathbf{x}_i) & \cdots & f_{pp}(\mathbf{x}_i) \\ \vdots & \vdots & \vdots & \vdots & \vdots & \vdots & & \\ f_0(\mathbf{x}_n) & f_1(\mathbf{x}_n) & \cdots & f_p(\mathbf{x}_n) & f_{11}(\mathbf{x}_n) & f_{12}(\mathbf{x}_n) & \cdots & f_{pp}(\mathbf{x}_n) \end{pmatrix} \in \mathcal{R}^{n \times (p+1)(p+2)/2},$$

where for $i = 1, \ldots, n$, $\mathbf{x}_i := (x_{i1}, \ldots, x_{ip})^\top \in \mathbf{I}^p$, and for $k, \ell = 1, \ldots, p$ $f_{k\ell}(\mathbf{x}) := f_k(\mathbf{x}) f_\ell(\mathbf{x})$, for $\mathbf{x} \in \mathbf{I}^p$. By referring to the work of Pruscha [14], pp. 115–117, there exist an $n_0 \geq 1$, such that $rank(\mathbf{X}_n) = (p+1)(p+2)/2$ for all $n \geq n_0$. For our purpose we need to represent the model using vector and matrix notation. Let $\mathbf{Y}_{nn} := (Y_{n1}, \ldots, Y_{nn})$, $\varepsilon_{nn} := (\varepsilon_{n1}, \ldots, \varepsilon_{nn})^\top$ be the vector of observations and random errors, respectively. Then Model 3 can be equivalently written as $\mathbf{Y}_{nn} = \mathbf{X}_n \beta + \varepsilon_{nn}$, where $Cov(\varepsilon_{nn}) = \mathbf{\Sigma} := (\sigma_{ij})_{i=1,j=1}^{n,n}$ is an $n \times n$ symmetric matrix. As an example Schabenberger and Gotway [15], pp. 320-321 studied a spatial regression model with correlated observations. They considered $\mathbf{\Sigma} = \sigma^2 \Psi = \sigma^2 (\psi_{ij})_{i=1,j=1}^{n,n}$, where $\psi_{ij} = \exp\{-\|\mathbf{x}_i - \mathbf{x}_j\|\}$, for $i, j = 1, \ldots, n$.

If the covariance matrix $\mathbf{\Sigma}$ is positive definite ($\mathbf{\Sigma} > 0$), then there exists $\mathbf{\Sigma}^{-1/2}$ such that $\mathbf{\Sigma}^{-1/2}\mathbf{\Sigma}^{-1/2} = \mathbf{\Sigma}^{-1}$. By using the linear transformation represented by $\mathbf{\Sigma}^{-1/2}$, the asymptotic model $\mathbf{Y}_{nn} = \mathbf{X}_n \beta^{(n)} + \varepsilon_{nn}$ can be transformed to $\mathbf{Y}_{nn}^* = \mathbf{X}_n^* \beta^{(n)} + \varepsilon_{nn}^*$, where $\mathbf{Y}_{nn}^* = \mathbf{\Sigma}^{-1/2} \mathbf{Y}_{nn}$, $\mathbf{X}_n^* = \mathbf{\Sigma}^{-1/2} \mathbf{X}_n$, and $\varepsilon_{nn}^* = \mathbf{\Sigma}^{-1/2} \varepsilon_{nn}$. The transformed model has the property $\mathbf{E}(\varepsilon_{nn}^*) = \mathbf{E}(\mathbf{\Sigma}^{-1/2}\varepsilon_{nn}) = \mathbf{0}$ and $Cov(\varepsilon_{nn}^*) = Cov(\mathbf{\Sigma}^{-1/2}\varepsilon_{nn}) = \mathbf{I}_n$. Hence the estimator for β is immediately obtained by applying the least squares principle (cf. [1, 2, 18]) which is given by

$$\widehat{\beta}^{(n)} = (\mathbf{X}_n^\top \mathbf{\Sigma}^{-1} \mathbf{X}_n)^{-1} \mathbf{X}_n^\top \mathbf{\Sigma}^{-1} \mathbf{Y}_{nn}.$$

Moreover, by the plug-in principle we get the least squares estimator of matrix $\mathbf{B}^{(n)}$. That is

$$\widehat{\mathbf{B}}^{(n)} := \begin{pmatrix} \hat{\beta}_{11}^{(n)} & 1/2\hat{\beta}_{12}^{(n)} & \cdots & 1/2\hat{\beta}_{1p}^{(n)} \\ 1/2\hat{\beta}_{12}^{(n)} & \hat{\beta}_{22}^{(n)} & \cdots & 1/2\hat{\beta}_{2p}^{(n)} \\ \vdots & \vdots & \ddots & \vdots \\ 1/2\hat{\beta}_{1p}^{(n)} & 1/2\hat{\beta}_{2p}^{(n)} & \cdots & \hat{\beta}_{pp}^{(n)} \end{pmatrix},$$

where the entries of $\widehat{\mathbf{B}}^{(n)}$ is obtained from the corresponding the elements of $\widehat{\beta}^{(n)}$. Furthermore, by applying the result presented in Pruscha [14], pp. 115–117, we get the limit distribution of

$Vech(\widehat{\mathbf{B}}^{(n)})$. Suppose the components of ε^*_{nn} are independent then for $n \to \infty$, it holds

$$\sqrt{n}(Vech(\widehat{\mathbf{B}}^{(n)}) - Vech(\mathbf{B})) \Rightarrow N_{\frac{1}{2}p(p+1)}(\mathbf{O}, \mathbf{T}_p\mathbf{G}^{-1}\mathbf{T}_p^\top),$$

where $\mathbf{G} := \lim_{n\to\infty} \mathbf{X}_n^{*\top}\mathbf{X}_n^*$ and $\mathbf{T}_p$ is the linear operator represented by a $\frac{1}{2}p(p+1) \times \frac{1}{2}(p+1)(p+2)$ matrix defined by

$$\mathbf{T}_p := \left(O|diag(\mathbf{a}_p^\top, \mathbf{a}_{p-1}^\top, \ldots, \mathbf{a}_1^\top)\right),$$

where O is the $\frac{1}{2}p(p+1) \times (p+1)$ zero matrix, and $\mathbf{a}_p := (1, 1/2, 1/2, \ldots, 1/2)^\top \in \mathcal{R}^p$. Here and throughout this paper "$\Rightarrow$" stand for the weak convergence or convergence in distribution (cf. Billingsley [4]).

To get the asymptotic distribution of the sequence of the vector of eigenvalues of $\widehat{\mathbf{B}}^{(n)}$ we need further notations. It is worth mentioning that the Euclidean space $\mathcal{R}^{p\times p}$ is furnished with an inner product denoted by $\langle \cdot, \cdots \rangle_{\mathcal{R}^{p\times p}}$, defined by $\langle \mathbf{A}, \mathbf{C}\rangle_{\mathcal{R}^{p\times p}} := trace(\mathbf{A}^\top\mathbf{C})$, and the corresponding norm is denoted by $\|\mathbf{A}\|_{\mathcal{R}^{p\times p}} := \sqrt{trace(\mathbf{A}^\top\mathbf{A})}$, for every $\mathbf{A}$ and $\mathbf{C}$ in $\mathcal{R}^{p\times p}$. A closed ball in $\mathcal{R}^{p\times p}$ with radius $r > 0$ and a center $\mathbf{A}$ is denoted by $\mathcal{N}_r(\mathbf{A})$, defined by $\mathcal{N}_r(\mathbf{A}) := \{\mathbf{X} \in \mathcal{R}^{p\times p} : \|\mathbf{X} - \mathbf{A}\|_{\mathcal{R}^{p\times p}} \le r\}$. Note that "$Vech$" denotes the well-known Vech operator defined in the literatures of matrix algebra, see e.g. Harville [7], pp. 340–343.

Now we are ready to state our main result giving the asymptotic distribution of the vector of the eigenvalues of $\widehat{\mathbf{B}}^{(n)}$. See also [3, 5].

Theorem 2.1. ([3, 5]) Suppose that the components of $\mathbf{\Sigma}^{-1/2}\varepsilon_{nn}$ are independent. Let λ_s be a simple eigenvalue of $\mathbf{B}$ with associated normalized eigenvector $\mathbf{u}_s := (u_{s1}, \ldots, u_{sp})^\top$. That is $\mathbf{B}\mathbf{u}_s = \lambda_s\mathbf{u}_s$ and $\mathbf{u}_j^\top\mathbf{u}_k = \delta_{jk}$, for $s, i, j = 1, \ldots, p$, where δ_{jk} is the Kronecker delta. Let $\hat{\lambda}_s^{(n)}$ be the eigenvalue of $\widehat{\mathbf{B}}^{(n)}$ with normalized eigenvector $\widehat{\mathbf{u}}_s^{(n)} := (\hat{u}_{s1}^{(n)}, \ldots, \hat{u}_{sp}^{(n)})^\top$. That is $\widehat{\mathbf{B}}^{(n)}\widehat{\mathbf{u}}_s^{(n)} = \hat{\lambda}_s^{(n)}\widehat{\mathbf{u}}_s^{(n)}$, for $s = 1, \ldots, p$. Furthermore, let $\mathbf{H}$ be a $\frac{1}{2}p(p+1) \times p$ real matrix (Hessian matrix) defined by

$$\mathbf{H} := \begin{pmatrix} Vech^\top(\mathbf{u}_1\mathbf{u}_1^\top - \mathbf{D}_1) \\ Vech^\top(\mathbf{u}_2\mathbf{u}_2^\top - \mathbf{D}_2) \\ \vdots \\ Vech^\top(\mathbf{u}_p\mathbf{u}_p^\top - \mathbf{D}_p) \end{pmatrix}^\top \in \mathcal{R}^{\frac{1}{2}p(p+1)\times p},$$

where for $s = 1, \ldots, p$, $\mathbf{D}_s := diag(u_{s1}^2, \ldots, u_{sp}^2)$, and $\mathbf{W}_p := \mathbf{H}^\top\mathbf{T}_p\mathbf{G}^{-1}\mathbf{T}_p^\top\mathbf{H}$. Then for $n \to \infty$, it holds true

$$\sqrt{n}(\widehat{\lambda}^{(n)} - \lambda) \Rightarrow N_p(\mathbf{0}, \mathbf{W}_p).$$

Proof: By Theorem 7 in Magnus and Neudecker [9], there exist continuously differentiable eigenvalue and eigenvector functions λ and $\mathbf{u}$ defined on the open ball $\mathcal{N}_r(\mathbf{B}) \cap Sym(p)$ centered at the unknown parameter $\mathbf{B}$, such that for every $\mathbf{X} \in \mathcal{N}_r(\mathbf{B}) \cap Sym(p)$, it holds $\mathbf{X}\mathbf{u}(\mathbf{X}) = \lambda(\mathbf{X})\mathbf{u}(\mathbf{X})$. In particular, we have $\mathbf{B}\mathbf{u}(\mathbf{B}) = \lambda(\mathbf{B})\mathbf{u}(\mathbf{B})$ and $\widehat{\mathbf{B}}^{(n)}\widehat{\mathbf{u}}^{(n)} = \widehat{\lambda}^{(n)}\widehat{\mathbf{u}}^{(n)}$. Next for every $\mathbf{t} := (t_1, \ldots, t_p)^\top \in \mathcal{R}^p$, we define a function $F_\mathbf{t}$ on the open ball $\mathcal{N}_r(\mathbf{B}) \cap Sym(p)$, such that for every $\mathbf{X} \in \mathcal{N}_r(\mathbf{B}) \cap Sym(p)$, $F_\mathbf{t}(\mathbf{X}) := \mathbf{t}^\top\lambda(\mathbf{X})$. By Taylor expansion, we have an approximation of $F_\mathbf{t}(\mathbf{X})$ for any $\mathbf{X}$ lies in the ball $\mathbf{X} \in \mathcal{N}_r(\mathbf{B}) \cap Sym(p)$. Especially we have

$$\sqrt{n}(F_\mathbf{t}(\widehat{\mathbf{B}}^{(n)}) - F_\mathbf{t}(\mathbf{B})) = \sqrt{n}Vech^\top(\widehat{\mathbf{B}}^{(n)} - \mathbf{B})\frac{\partial F_\mathbf{t}(\mathbf{B})}{\partial Vech(\mathbf{X})} + R_{\widehat{\mathbf{B}}^{(n)}}, \tag{4}$$

where $R_{\widehat{\mathbf{B}}^{(n)}} := \|\sqrt{n}Vech(\widehat{\mathbf{B}}^{(n)} - \mathbf{B})\| r(Vech(\widehat{\mathbf{B}}^{(n)} - \mathbf{B}))$. We notice that $R_{\widehat{\mathbf{B}}^{(n)}}$ immediately converges to zero stochastically by the consistence of $\widehat{\mathbf{B}}^{(n)}$ to $\mathbf{B}$. Next by substituting

$$\begin{aligned}\frac{\partial F_{\mathbf{t}}(\mathbf{B})}{\partial Vech(\mathbf{X})} &= Vech\left(\frac{\partial F_{\mathbf{t}}(\mathbf{B})}{\partial \mathbf{X}}\right) = Vech\left(\frac{\sum_{s=1}^{p} t_s\lambda_s}{\partial \mathbf{X}}\right)\\ &= \sum_{s=1}^{p} t_s Vech\left(\frac{\partial \lambda_s}{\partial \mathbf{X}}\right) = \sum_{s=1}^{p} t_s Vech(2\mathbf{u}_s\mathbf{u}_s^\top - \mathbf{D}_s)\end{aligned}$$

to Equation 4 we finally get

$$\begin{aligned}\sqrt{n}(\widehat{\lambda}^{(n)} - \lambda) &= \sqrt{n}Vech^\top(\widehat{\mathbf{B}}^{(n)} - \mathbf{B})\sum_{s=1}^{p} t_s Vech(2\mathbf{u}_s\mathbf{u}_s^\top - \mathbf{D}_s) + R_{\widehat{\mathbf{B}}^{(n)}}\\ &= \sqrt{n}\mathbf{t}^\top\mathbf{H}^\top Vech(\widehat{\mathbf{B}}^{(n)} - \mathbf{B}) + R_{\widehat{\mathbf{B}}^{(n)}}.\end{aligned}$$

Since $R_{\widehat{\mathbf{B}}^{(n)}}$ converges to zero, and $\sqrt{n}Vech(\widehat{\mathbf{B}}^{(n)} - \mathbf{B})$ converges to $N_{\frac{1}{2}p(p+1)}(0, \mathbf{T}_p\mathbf{G}^{-1}\mathbf{T}_p^\top)$, then by applying the Cramér-Wald technique, the assertion follows.

Remark 2.1. In the application the covariance matrix is estimated by a positive definite matrix

$$\widehat{\mathbf{\Sigma}}^{(n)} := \frac{1}{n}(\mathbf{I}_n - \mathbf{X}_n(\mathbf{X}_n^\top\mathbf{X}_n)^{-1}\mathbf{X}_n^\top)\mathbf{Y}_{nn}\mathbf{Y}_{nn}^\top(\mathbf{I}_n - \mathbf{X}_n(\mathbf{X}_n^\top\mathbf{X}_n)^{-1}\mathbf{X}_n^\top),$$

where $\mathbf{I}_n$ is the $n\times n$ identity matrix. The Hessian matrix is estimated by the following consistent estimator

$$\widehat{\mathbf{H}}^{(n)} := \begin{pmatrix} Vech^\top(\widehat{\mathbf{u}}_1^{(n)}\widehat{\mathbf{u}}_1^{(n)\top} - \widehat{\mathbf{D}}_1^{(n)})\\ Vech^\top(\widehat{\mathbf{u}}_2^{(n)}\widehat{\mathbf{u}}_2^{(n)\top} - \widehat{\mathbf{D}}_2^{(n)})\\ \vdots\\ Vech^\top(\widehat{\mathbf{u}}_p^{(n)}\widehat{\mathbf{u}}_p^{(n)\top} - \widehat{\mathbf{D}}_p^{(n)})\end{pmatrix}^\top \in \mathcal{R}^{\frac{1}{2}p(p+1)\times p}.$$

Hence by the plug-in principle, $\mathbf{W}_p$ is estimated by $\widehat{\mathbf{W}}_p^{(n)} := \widehat{\mathbf{H}}^{(n)\top}\mathbf{T}_p\mathbf{G}^{-1}\mathbf{T}_p^\top\widehat{\mathbf{H}}^{(n)}$.

Remark 2.2. In the case of the covariance structure $\mathbf{\Sigma} = \sigma^2\mathbf{\Psi}$ defined in [15], a consistent estimator for σ^2 proposed in [1, 2] can be applied. That is

$$\widehat{\sigma}^2_{(n)} := \frac{\mathbf{Y}_{nn}^*(\mathbf{I}_n - \mathbf{X}_n^*(\mathbf{X}_n^{*\top}\mathbf{X}_n^*)^{-1}\mathbf{X}_n^{*\top})\mathbf{Y}_{nn}^*}{n - \frac{1}{2}(p+1)(p+2)}.$$

As an immediate implication of the multivariate central limit theorem (Pruscha [14]) we have

$$\sqrt{n}(\widehat{\lambda}^{(n)} - \lambda)^\top(\widehat{\mathbf{W}}_p^{(n)})^{-1}\sqrt{n}(\widehat{\lambda}^{(n)} - \lambda) \Rightarrow \chi^2(p),\ for\ n \to \infty,$$

where $\chi^2(p)$ is the random variable having a chi-square distribution with p degrees of freedom. Hence, the $(1-\alpha)\times 100\%$ confidence region of λ is given by

$$\mathcal{R}_\lambda := \left\{\mathbf{x} \in \mathcal{R}^p :\ \sqrt{n}(\widehat{\lambda}^{(n)} - \mathbf{x})^\top(\widehat{\mathbf{W}}_p^{(n)})^{-1}\sqrt{n}(\widehat{\lambda}^{(n)} - \mathbf{x}) \le \chi^2_{1-\alpha}(p)\right\}, \tag{5}$$

where $\chi^2_{1-\alpha}(p)$ is the $(1-\alpha)$ quantile of the chi-square distribution with p degrees of freedom. One is interested not only on the simultaneous confidence region for the vector λ but also on the partial confidence interval for each component of λ. For $s = 1, \ldots, p$, let $\mathbf{1}_s$ be the unit vector in $\mathcal{R}^p$ whose s-th component is 1 and the remainder is 0. Then by Theorem 2.1 we get

$$\frac{\sqrt{n}\mathbf{1}_s^\top(\widehat{\lambda}^{(n)} - \lambda)}{\sqrt{\mathbf{1}_s^\top \widehat{\mathbf{W}}_p^{(n)} \mathbf{1}_s}} \Rightarrow N(0, 1),$$

giving us the partial $(1-\alpha) \times 100\%$ confidence interval for the component λ_s of λ

$$\left(\widehat{\lambda}_s^{(n)} - z_{1-\alpha/2}\frac{1}{\sqrt{n}}\sqrt{\mathbf{1}_s^\top \widehat{\mathbf{W}}_p^{(n)} \mathbf{1}_s},\ \widehat{\lambda}_s^{(n)} + z_{1-\alpha/2}\frac{1}{\sqrt{n}}\sqrt{\mathbf{1}_s^\top \widehat{\mathbf{W}}_p^{(n)} \mathbf{1}_s}\right), \tag{6}$$

where $z_{1-\alpha}$ is the $(1-\alpha)$-th quantile of the standard normal distribution.

Algorithm for constructing the confidence region or confidence interval for the eigenvalue of $\mathbf{B}$ is presented bellow.

(i) Compute $\widehat{\mathbf{\Sigma}}^{(n)}$;

(ii) Compute $\widehat{\beta}^{(n)}$;

(iii) Compute the estimated moel matrix $\widehat{\mathbf{B}}^{(n)}$;

(iv) Compute the eigenvalue vector $\widehat{\lambda}^{(n)}$;

(v) Compute $\widehat{\mathbf{W}}_p^{(n)}$;

(vi) Construct the confidence region by using (5) and the confidence interval by using (6).

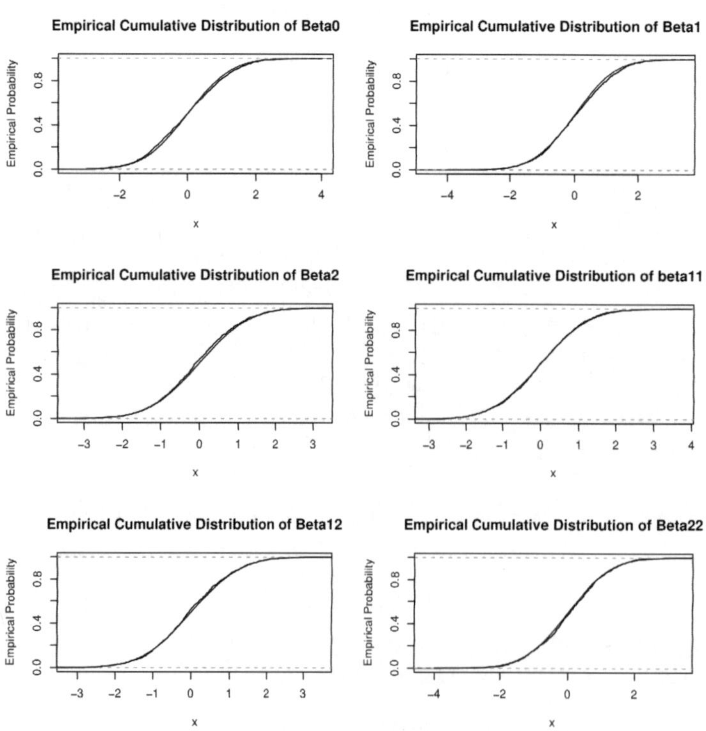

Figure 1. The empirical distribution functions of $\widehat{\beta}_0^{(n)}$, $\widehat{\beta}_1^{(n)}$, $\widehat{\beta}_2^{(n)}$, $\widehat{\beta}_{11}^{(n)}$, $\widehat{\beta}_{12}^{(n)}$ and $\widehat{\beta}_{22}^{(n)}$.

3. Simulation study

We develop simulations in investigating the behavior of the finite sample distribution of the pivotal quantities defined in the preceding section. The sample is generated from a second-order RSM with the experimental design given by $n \times n$ regular lattice on the rectangle $[0,1] \times [0,1]$,

$$Y_{\ell k} = \beta_0 + \beta_1 \ell/n + \beta_2 k/n + \beta_{11}(\ell/n)^2 + \beta_{12}\ell k/(n^2) + \beta_{22}(k/n)^2 + \varepsilon_{\ell k}, \ 1 \le \ell, k \le n.$$

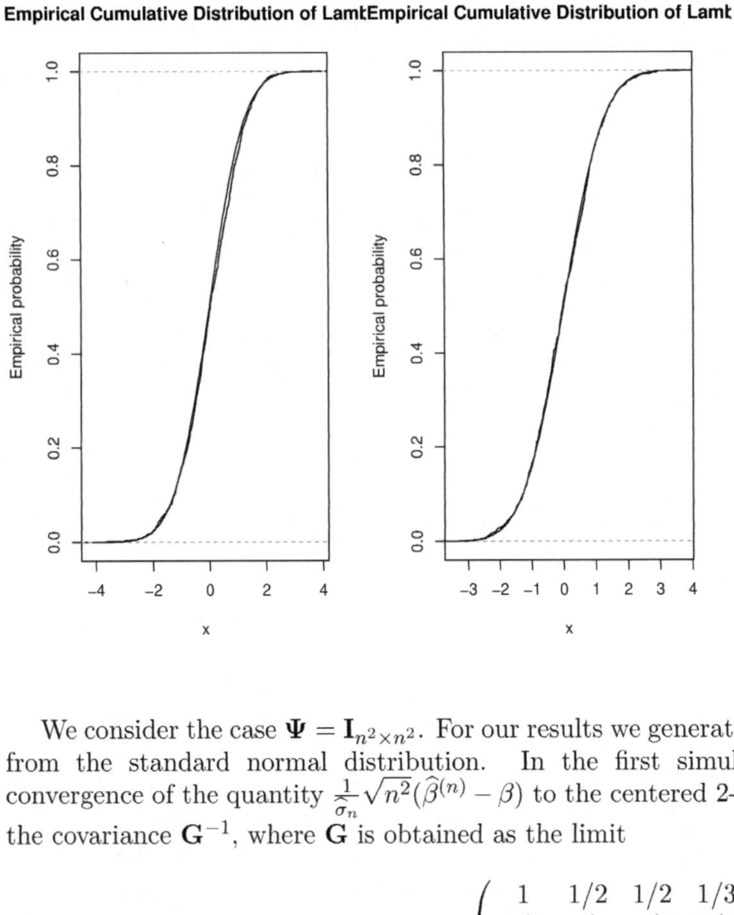

Figure 2. The empirical distribution functions of λ_1 (left panel) and λ_2 (right panel).

We consider the case $\mathbf{\Psi} = \mathbf{I}_{n^2 \times n^2}$. For our results we generated the error terms independently from the standard normal distribution. In the first simulation we investigate the weak convergence of the quantity $\frac{1}{\widehat{\sigma}_n}\sqrt{n^2}(\widehat{\beta}^{(n)} - \beta)$ to the centered 2-variate normal distribution with the covariance $\mathbf{G}^{-1}$, where $\mathbf{G}$ is obtained as the limit

$$\mathbf{G} = \lim_{n\to\infty} \frac{1}{n^2}\mathbf{X}_n^\top \mathbf{X}_n = \begin{pmatrix} 1 & 1/2 & 1/2 & 1/3 & 1/4 & 1/3 \\ 1/2 & 1/3 & 1/4 & 1/4 & 1/6 & 1/6 \\ 1/2 & 1/4 & 1/3 & 1/6 & 1/6 & 1/4 \\ 1/3 & 1/4 & 1/6 & 1/5 & 1/8 & 1/9 \\ 1/4 & 1/6 & 1/6 & 1/8 & 1/9 & 1/8 \\ 1/3 & 1/6 & 1/4 & 1/9 & 1/8 & 1/5 \end{pmatrix}.$$

Figure 1 exhibits the graphs of the empirical distribution function of $\widehat{\beta}_0^{(n)}$, $\widehat{\beta}_1^{(n)}$, $\widehat{\beta}_2^{(n)}$, $\widehat{\beta}_{11}^{(n)}$, $\widehat{\beta}_{12}^{(n)}$ and $\widehat{\beta}_{22}^{(n)}$ indicated by dotted line, whereas the graph of the CDF of the standard normal distribution is represented as smooth line. In the second case we consider the finite sample distribution of the quantity $\frac{\sqrt{n}\mathbf{1}_s^\top(\widehat{\lambda}^{(n)}-\lambda)}{\widehat{\sigma}_{(n)}\sqrt{\mathbf{1}_s{}^\top \widehat{\Sigma}_p^{(n)}\mathbf{1}_s}}$, for $s = 1, 2$, which is shown to converge to the standard normal distribution. The simulation result is presented in Figure 2. Next in Figure 3

Journal of Physics: Conference Series **725** (2016) 012002 doi:10.1088/1742-6596/725/1/012002

we present the empirical distribution function os the quantity

$$\mathbf{V}_n := \frac{1}{\widehat{\sigma}_{(n)}}\sqrt{n}(\widehat{\lambda}^{(n)} - \lambda)^\top(\widehat{\Sigma}_p^{(n)})^{-1}\frac{1}{\widehat{\sigma}_{(n)}}\sqrt{n}(\widehat{\lambda}^{(n)} - \lambda).$$

The dotted line is the empirical distribution function of $\mathbf{V}_n$, whereas the smooth line is the empirical distribution function of a chi-square distribution with 2 degrees of freedom.

Based on the simulation result it can be concluded that our approach gives us a reasonable way to approximate the confidence region of the eigenvalues of $\widehat{\mathbf{B}}^{(n)}$.

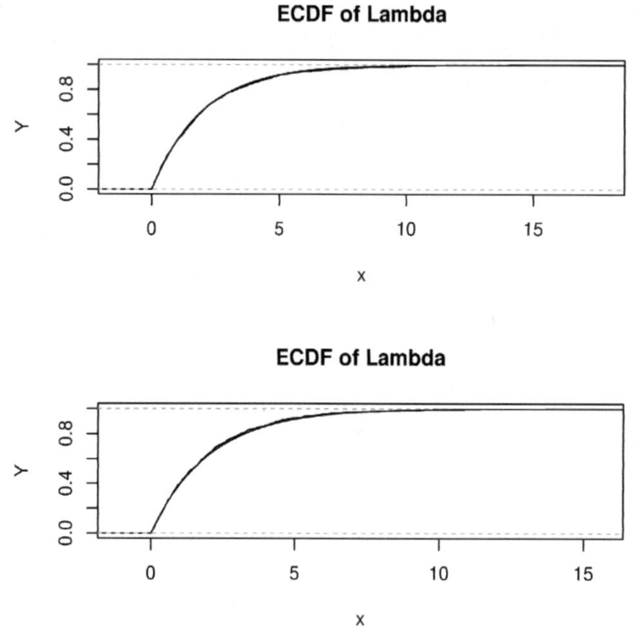

Figure 3. The empirical distribution functions of $\mathbf{V}_n$ (dotted line) for $n = 25$ and 35.

4. Application to a mining data

We consider a data obtained from a mining industry in Southeast Sulawesi. The data is the percentage of Co (Cobalt) recorded over 98 points of drilling bore with 14 equidistance columns running south to north and 7 equidistance rows running west to east. By performing model-check using the partial sums technique proposed in Somayasa and et. al. [16] a second-order polynomial model with isotropic observation is appropriate for representing the model.

We assume a covariance structure as in [15]. After transforming the experimental region to the unit rectangle $[0, 1] \times [0, 1]$, we get

$$\widehat{\mathbf{b}}^{(n)} = (-0.847, 0.611)^\top,\ \widehat{\mathbf{B}}^{(n)} = \begin{pmatrix} -1.146 & +4.949 \\ +4.949 & -5.579 \end{pmatrix},\ \widehat{\mathbf{x}}_0 = (0.047, 0.096)^\top.$$

The eigenvalue and the associated eigenvector of $\widehat{\mathbf{B}}^{(n)}$ are given by

$$\widehat{\lambda}^{(n)} = (2.060, -8.785)^\top, \widehat{\mathbf{u}}_1^{(n)} = (-0.839, -0.544)^\top, \widehat{\mathbf{u}}_2^{(n)} = (0.544, -0.839)^\top.$$

By the transformation

Journal of Physics: Conference Series **725** (2016) 012002 doi:10.1088/1742-6596/725/1/012002

Table 1. Percentages of Co (Cobalt) observed over 96 points (x, y) of bore.

x	y	Co(%)	x	y	Co(%)	x	y	Co(%)	x	y	Co(%)
-5825	-6000	0.07	-5825	-5900	0.06	-5750	-5825	0.04	-5675	-5750	0.07
-5800	-6000	0.06	-5800	-5900	0.06	-5725	-5825	0.04	-5825	-5725	0.05
-5775	-6000	0.06	-5775	-5900	0.05	-5700	-5825	0.06	-5800	-5725	0.06
-5750	-6000	0.05	-5750	-5900	0.05	-5675	-5825	0.08	-5775	-5725	0.05
-5725	-6000	0.08	-5725	-5900	0.05	-5825	-5800	0.06	-5750	-5725	0.05
-5700	-6000	0.07	-5700	-5900	0.10	-5800	-5800	0.05	-5725	-5725	0.05
-5675	-6000	0.04	-5675	-5900	0.14	-5775	-5800	0.04	-5700	-5725	0.04
-5825	-5975	0.07	-5825	-5875	0.05	-5750	-5800	0.04	-5675	-5725	0.06
-5800	-5975	0.08	-5800	-5875	0.05	-5725	-5800	0.13	-5825	-5700	0.04
-5775	-5975	0.09	-5775	-5875	0.05	-5700	-5800	0.13	-5800	-5700	0.05
-5750	-5975	0.08	-5750	-5875	0.07	-5675	-5800	0.10	-5775	-5700	0.07
-5725	-5975	0.10	-5725	-5875	0.08	-5825	-5775	0.09	-5750	-5700	0.07
-5700	-5975	0.11	-5700	-5875	0.05	-5800	-5775	0.08	-5725	-5700	0.06
-5675	-5975	0.13	-5675	-5875	0.06	-5775	-5775	0.07	-5700	-5700	0.07
-5825	-5950	0.06	-5825	-5850	0.06	-5750	-5775	0.06	-5675	-5700	0.05
-5800	-5950	0.09	-5800	-5850	0.06	-5725	-5775	0.06	-5825	-5675	0.05
-5775	-5950	0.12	-5775	-5850	0.07	-5700	-5775	0.07	-5800	-5675	0.05
-5750	-5950	0.08	-5750	-5850	0.07	-5675	-5775	0.06	-5775	-5675	0.05
-5725	-5950	0.10	-5725	-5850	0.14	-5825	-5750	0.06	-5750	-5675	0.05
-5700	-5950	0.12	-5700	-5850	0.09	-5800	-5750	0.05	-5725	-5675	0.06
-5675	-5950	0.08	-5675	-5850	0.05	-5775	-5750	0.04	-5700	-5675	0.05
-5825	-5925	0.05	-5825	-5825	0.05	-5750	-5750	0.05	-5675	-5675	0.05
-5800	-5925	0.05	-5800	-5825	0.04	-5725	-5750	0.05	-5750	-5925	0.06
-5775	-5925	0.07	-5775	-5825	0.05	-5700	-5750	0.06	-5725	-5925	0.06
-5700	-5925	0.07	-5675	-5925	0.07						

Source of Data: PT. Aneka Tambang Tbk. Pomalaa, available in Tahir [19].

$$\begin{pmatrix} w_1 \\ w_2 \end{pmatrix} = \begin{pmatrix} -0.839 & -0.544 \\ +0.544 & -0.839 \end{pmatrix} \begin{pmatrix} x_1 - 0.047 \\ x_2 - 0.096 \end{pmatrix},$$

we obtain the canonical form of the fitted model as $\widehat{Y}^{(n)} = 0.947 + 2.060w_1^2 - 8.785w_2^2$. Since the estimated eigenvalues have different signs, we can infer that the estimated stationary point $\widehat{\mathbf{x}}_0$ seems to be a saddle point. We get the estimated Hessian matrix and the estimated covariance matrix of $\widehat{\lambda}^{(n)}$ as

$$\widehat{\mathbf{H}}^{(n)} = \begin{pmatrix} +0.704 & +0.913 & +0.296 \\ +0.296 & -0.913 & +0.704 \end{pmatrix}, \ \widehat{\Sigma}_p^{(n)} = \begin{pmatrix} 135.357 & 45.207 \\ 45.207 & 135.357 \end{pmatrix}.$$

Hence the approximated $(1-\alpha) \times 100\&$ confidence region for λ is given by

$$\left\{ \begin{pmatrix} \lambda_1 \\ \lambda_2 \end{pmatrix} \in \mathcal{R}^2 : \begin{pmatrix} +2.060 - \lambda_1 \\ -8.785 - \lambda_2 \end{pmatrix}^{\top} \begin{pmatrix} +0.008 & -0.003 \\ -0.003 & +0.008 \end{pmatrix} \begin{pmatrix} +2.060 - \lambda_1 \\ -8.785 - \lambda_2 \end{pmatrix} \leq \frac{1}{1600} \chi^2_{1-\alpha}(2) \right\}.$$

The last region is actually an ellipsoid on two dimensional region. To see this let us define a new coordinate system $(\lambda_1', \lambda_2')^\top$ according to the transformation formula

$$\begin{pmatrix} \lambda_1 \\ \lambda_2 \end{pmatrix} = \begin{pmatrix} -\frac{1}{2}\sqrt{2} & -\frac{1}{2}\sqrt{2} \\ +\frac{1}{2}\sqrt{2} & -\frac{1}{2}\sqrt{2} \end{pmatrix} \begin{pmatrix} \lambda_1' \\ \lambda_2' \end{pmatrix} + \begin{pmatrix} +2.060 \\ -8.785 \end{pmatrix}.$$

Then by substituting λ_1 and λ_2 then we get the new confidence region in term of λ_1' and λ_2':

$$\left\{ \begin{pmatrix} \lambda_1' \\ \lambda_2' \end{pmatrix} \in \mathcal{R}^2 : \left(\frac{\lambda_1'}{\sqrt{\frac{1}{17.6}\chi^2_{1-\alpha}(2)}} \right) + \left(\frac{\lambda_2'}{\sqrt{\frac{1}{8}\chi^2_{1-\alpha}(2)}} \right) \leq 1 \right\}.$$

The computation result gives us information that the stationary point of the process is set on the point $(0.047, 0.096)$ which lies on the experimental region. Unfortunately the stationary point associated to a saddle point. Therefore we can not determine whether on that point the process attains a maximum or a minimum value.

5. Conclusion

We have developed an asymptotic method for constructing the confidence region as well as confidence interval for the eigenvalues of the model matrix in normal second-order response surface model with spatially correlated observations. From the practical view point the computation of the derived formula is easily executed so that it seems to be very helpful for the practitioner. Given a data observed from a second-order RSM the optimum property of the stationary point can be immediately identified by conducting a computation according the algorithm presented at the end of Section 2. The stationary point of the percentage of Co data which lies on the point $(0.047, 0.096)$ is a saddle point.

Acknowledgement

The authors wishes to thank the Indonesian Ministry of Research and Technology and Higher Education for the financial support through the SAME 2015 program. A special thank is also dedicated to the Institut fuer Stochastik of Karlsruhe Institut fuer Technologie for hospitality.

References

[1] Arnold S F 1981 *The Theory of Linear Models and Multivarite Analysis* (New York: John Wiley & Sons)
[2] Arnold S F 1984 Journal of the American Statistical Association 75**(372)** 890–4.
[3] Bisgaard S and Ankenman B 1996 Technometrics 38 **(3)** 238-46.
[4] Billingsley P 1999 *Convergence of Probability Measures* (New York: John Wiley & Sons)
[5] Carter W H, Chincilli V M and Campbell E D 1990 Technometric **32** 425–35.
[6] Cressie Noel A C 1991 Tatistics for Spatial Data (New York: John Wiley & Sons)
[7] Harville D A 1997 Matrix Algebra from a Statistician's Perspective (New York: Springer-Verlag)
[8] Khuri A I and Cornell J A 1996 Response Surfaces, Design and Analysis (2nd. edition, revised and expanded) (New York: Marcel Dekker)
[9] Magnus J R and Neudecker H 1988 Matrix Differential Calculus with Applications in Statistics and Econometrics (New York: John Wiley & Sons).
[10] Myers R H and Kuhri A I 1979 Commun Stat: Theory and Methods **8** 1359–76.
[11] Myers R H and Montgomery D C 1996 Response Surface Methodology (New York: John Wiley & Sons)
[12] Myers R H and Montgomery D C 2002 Response Surface Methodology: Process and Product Optimization Using Designed Experiments, 2nd editiion (New York: John Wiley & Sons)
[13] Paul S and Khuri A I 2000 Commun Stat: Theory Methods **29** 2181-200.
[14] Pruscha H 2000 Vorlesungen ber Mathematische Statistik (Stuttgart: B. G. Teubner)
[15] Schabenberger O and Gotway C A 2005 Statistical Methods for Spatial Data Analysis (Florida: Chapman)
[16] Somayasa W, Ruslan, Cahyono E and Engkoimani Fareast Journal of Mathematical Sciences 96**(8)** 933–66.
[17] Somayasa W 2009 Paradigma 13 **(2)** 105–16.
[18] Stapleton J H 1995 *Linear Statistical Models* (New York: John Wiley & Sons)
[19] Tahir M 2010 *Prdiction of the Amount of Nickel Deposit Based on the Result of dDriling Bores on Several Points* (Kendari: Halu Oleo University)

Journal of Physics: Conference Series **725** (2016) 012003 doi:10.1088/1742-6596/725/1/012003

Comparing tests appear in model-check for normal regression with spatially correlated observations

Wayan Somayasa and Gusti A. Wibawa

Department of Mathematics, Halu Oleo University, Kendari 93232, Indonesia

E-mail: `wayan.somayasa@uho.ac.id` and `gustiadhi.wibawa@uho.ac.id`

Abstract. The problem of investigating the appropriateness of an assumed model in regression analysis was traditionally handled by means of F test under independent observations. In this work we propose a more modern method based on the so-called set-indexed partial sums processes of the least squares residuals of the observations. We consider throughout this work univariate and multivariate regression models with spatially correlated observations, which are frequently encountered in the statistical modelling in geosciences as well as in mining. The decision is drawn by performing asymptotic test of statistical hypothesis based on the Kolmogorov-Smirnov and Cramér-von Misses functionals of the processes. We compare the two tests by investigating the power functions of the test. The finite sample size behavior of the tests are studied by simulating the empirical probability of rejections of H_0. It is shown that for univariate model the KS test seems to be more powerful. Conversely the Cramér-von Mises test tends to be more powerful than the KS test in the multivariate case.

1. Introduction

Checking the appropriateness of an assumed model becomes an important step in a serious empirical statistical model building in order to obtain a suitable model. By this reason the problem of model check or goodness-of-fit in regression analysis has obtained much attention. Classical F test studied in standard text book is frequently applied in the practice, see Graybill [10] and Seber and Lee [15] for the regression model with single response and Christensen [7] and Johnson and Wichern [11] for that with multiple responses. Asymptotic procedures have been proposed by authors such as Arnold [2] established F_n test for univariate model, whereas Arnold [3] and Zellner [22] proposed asymptotic method for multivariate model. The study for obtaining a better but more simple procedure was extended by Stute [19, 20] leading to the pioneering study using the empirical processes marked by the residuals.

The application of the set-indexed partial sums processes of the residuals in asymptotic model check for spatial regression was investigated in the work of Somayasa [16], Somayasa and et al [17, 18]. In these references the limit process was obtained by generalizing the geometric method proposed in Bischoff [5] and Bischoff and Somayasa [6]. The main idea of the study was motivated by the attempt of McNeill and Jandhyalla [12] and Xie and McNeill [21] to detect the existence of change in the regression functions by embedding the spatial residuals to stochastic process using the partial sums operator.

The partial sums approaches established in the literatures mentioned above were derived under a common assumption in that the observations are stochastically independent. However prediction method in geostatistics such as ordinary and universal kriging are mainly derived

Content from this work may be used under the terms of the Creative Commons Attribution 3.0 licence. Any further distribution of this work must maintain attribution to the author(s) and the title of the work, journal citation and DOI.
Published under licence by IOP Publishing Ltd

under spatially correlated observations, see e.g. Cressie [8]. Kriging for isotropic observations was studied in Schabenberger and Gotway [14]. By this reason in this paper we study the application of the set-indexed partial sums technique in investigating the correctness of the assumed model when the correlation between the observations are inherent. However the underlying distribution model of the sample is still assumed to be normal. This normality assumption is important in order to keep the independence among the observations holds true, see Section 2. The limit process for univariate and multivariate case are established in Section 3. Simulation study for comparing the finite sample power functions of the tests is presented in Section 4. The paper is closed with conclusion and some remark for future research, see Section 5.

2. Model definition and estimation

We consider univariate and multivariate models with correlated observations. In univariate case the response consists of one measurement only, whereas in multivariate case two or more variables are measured simultaneously on each individual or point. That why the analysis must be carried out by taking into account not only the correlation between the observations (points) but also among the response variables in each point. That why the mathematical derivation of the method becomes more complicated.

2.1. Univariate case

Throughout this paper we consider second-order stationary Gaussian spatial process

$$Y := \{Y(\mathbf{t}) :\ \mathbf{t} := (t_1, \ldots, t_d)^\top \in \Pi_{i=1}^d [a_i, b_i] := [\mathbf{a}, \mathbf{b}]\},$$

with the property $\mathbf{E}(Y(\mathbf{t})) = g(\mathbf{t})$ for an unknown real valued function g on $[\mathbf{a}, \mathbf{b}]$ and $Cov(Y(\mathbf{t}_j), Y(\mathbf{t}_k)) = \sigma(\mathbf{t}_j - \mathbf{t}_k)$, for every $\mathbf{t}_j, \mathbf{t}_k \in [\mathbf{a}, \mathbf{b}]$. That is Y has the covariance function which depends only on the difference $(\mathbf{t}_j - \mathbf{t}_k)$. Then the process Y admits a decomposition

$$Y(\mathbf{t}) = g(\mathbf{t}) + \varepsilon(\mathbf{t}),\ \mathbf{t} \in [\mathbf{a}, \mathbf{b}], \tag{1}$$

where $\varepsilon(\mathbf{t})$ is the random error which is normally distributed with $\mathbf{E}(\varepsilon(\mathbf{t})) = 0$ and $Cov(\varepsilon(\mathbf{t}_j), \varepsilon(\mathbf{t}_k)) = \sigma(\mathbf{t}_j - \mathbf{t}_k)$. In the isotropic spatial regression model, the covariance function of Y is assumed to depend only on the Euclidean distance between two points. As for example Sachenberger and Gotway [14] assume a process with the covariance function given by $Cov(Y(\mathbf{t}_j), Y(\mathbf{t}_k)) = \sigma^2 \exp\{-\|\mathbf{t}_j - \mathbf{t}_k\|\}$, for any $\mathbf{t}_j, \mathbf{t}_k \in [\mathbf{a}, \mathbf{b}]$, where σ^2 is unknown positive constant. In the references cited in Section 1 and also in the usual theory of regression analysis the covariance function of the process Y was assumed to be constant, i.e. $Cov(Y(\mathbf{t}_j), Y(\mathbf{t}_k)) = \sigma_{jk}\delta_{jk}$, where δ_{jk} is the Kronecker delta. In contrast, for our result we do not need such assumption.

Let $\mathbf{W} := [f_1, \ldots, f_p]$ be a subspace in $L_2(\lambda^d, [\mathbf{a}, \mathbf{b}])$, where $f_1, \ldots, f_p$ are known real valued regression functions on $[\mathbf{a}, \mathbf{b}]$ and $L_2(\lambda^d, [\mathbf{a}, \mathbf{b}])$ is the space of square integrable functions with respect to the Lebesgue measure λ^d on $[\mathbf{a}, \mathbf{b}]$. We aim to test the hypothesis of the form

$$H_0 : g \in \mathbf{W}\ versus\ H_1 : g \notin \mathbf{W}. \tag{2}$$

This means that under H_0 the unknown function g can be represented as a linear combination $g(\mathbf{t}) = \sum_{i=1}^p \beta_i f_i(\mathbf{t})$ for some unknown parameters $\beta_1, \ldots, \beta_p$, while under H_1 we consider a nonparametric model.

Suppose Model 1 is observed over an experimental design given by a regular lattice $\Xi_{n_1 \cdots n_d}$ with $n_1 \times \cdots \times n_d$ experimental condition (design points), where $n_1 \geq 1, \ldots, n_d \geq 1$. That is

$$\Xi_{n_1 \cdots n_d} := \{(t_{n_1 j_1}, \ldots, t_{n_d j_d}) :\ 1 \leq j_k \leq n_k,\ k = 1, \ldots, d\} \subset [\mathbf{a}, \mathbf{b}],$$

where for $k = 1, \ldots, d$, $t_{n_k j_k} := a_k + \frac{j_k}{n_k}(b_k - a_k)$. If we have an observation on every point of $\Xi_{n_1,\ldots,n_d}$ then we get an array of observed responses $\{Y_{j_1\cdots j_d} : 1 \le j_1 \le n_1, \ldots, 1 \le j_d \le n_d\}$ that satisfies the regression model

$$Y_{j_1\cdots j_d} = g_{j_1\cdots j_d} + \varepsilon_{j_1\cdots j_d},\ 1 \le j_1 \le n_1, \ldots, 1 \le j_d \le n_d,$$

where $Y_{j_1\cdots j_d}$ denotes the observation in the point $(t_{n_1 j_1}, \ldots, t_{n_d j_d})$. Furthermore, $Y_{j_1\cdots j_d}$ is normally distributed with $\mathbf{E}(Y_{j_1\cdots j_d}) = g_{j_1\cdots j_d}$ and $Cov(Y_{j_1\cdots j_d}, Y_{j_1'\cdots j_d'}) = \sigma(\mathbf{t}_{j_1\cdots j_d} - \mathbf{t}_{j_1'\cdots j_d'})$ for every point $\mathbf{t}_{j_1\cdots j_d} := (t_{n_1 j_1}, \ldots, t_{n_d j_d})$ and $\mathbf{t}_{j_1'\cdots j_d'} := (t_{n_1 j_1'}, \ldots, t_{n_d j_d'})$ in $\Xi_{n_1\cdots n_d}$.

In order to establish estimation procedure for the parameters of the model we need to describe the model in vectors and matrices forms. Let $\mathbf{Y}_{n_1\cdots n_d}$ be the $n_1\cdots n_d$-dimensional vector of random observation, $\mathbf{X}_{n_1\cdots n_d}$ be the $(n_1\cdots n_d)\times p$-dimensional design matrix, $\beta := (\beta_1, \ldots, \beta_p)^\top$ be p-dimensional vector of unknown parameters, and $\mathcal{E}_{n_1\cdots n_d}$ be the $n_1\cdots n_d$-dimensional random errors. Then the model under H_0 can also be written as

$$\mathbf{Y}_{n_1\cdots n_d} = \mathbf{X}_{n_1\cdots n_d}\beta + \mathcal{E}_{n_1\cdots n_d}, \tag{3}$$

where $\mathbf{Y}_{n_1\cdots n_d}$ has the $n_1\cdots n_d$ dimensional normal distribution with mean $\mathbf{X}_{n_1\cdots n_d}\beta$ and variance-covariance matrix given by $\sigma^2\mathbf{\Sigma}$, say, defined by

$$\mathbf{\Sigma} := \begin{pmatrix} \sigma(\mathbf{t}_{1\cdots 1} - \mathbf{t}_{1\cdots 1}) & \sigma(\mathbf{t}_{1\cdots 1} - \mathbf{t}_{21\cdots 1}) & \cdots & \sigma(\mathbf{t}_{1\cdots 1} - \mathbf{t}_{n_1\cdots n_d}) \\ \sigma(\mathbf{t}_{21\cdots 1} - \mathbf{t}_{1\cdots 1}) & \sigma(\mathbf{t}_{21\cdots 1} - \mathbf{t}_{21\cdots 1}) & \cdots & \sigma(\mathbf{t}_{21\cdots 1} - \mathbf{t}_{n_1\cdots n_d}) \\ \vdots & \vdots & \ddots & \vdots \\ \sigma(\mathbf{t}_{n_1\cdots n_d} - \mathbf{t}_{1\cdots 1}) & \sigma(\mathbf{t}_{n_1\cdots n_d} - \mathbf{t}_{21\cdots 1}) & \cdots & \sigma(\mathbf{t}_{n_1\cdots n_d} - \mathbf{t}_{n_1\cdots n_d}) \end{pmatrix}$$

which is considered as an unknown $(n_1\cdots n_d) \times (n_1\cdots n_d)$-dimensional matrix. In the case of spatial isotropy, we have

$$\mathbf{\Sigma} := \begin{pmatrix} \sigma(\|\mathbf{t}_{1\cdots 1} - \mathbf{t}_{1\cdots 1}\|) & \sigma(\|\mathbf{t}_{1\cdots 1} - \mathbf{t}_{21\cdots 1}\|) & \cdots & \sigma(\|\mathbf{t}_{1\cdots 1} - \mathbf{t}_{n_1\cdots n_d}\|) \\ \sigma(\|\mathbf{t}_{21\cdots 1} - \mathbf{t}_{1\cdots 1}\|) & \sigma(\|\mathbf{t}_{21\cdots 1} - \mathbf{t}_{21\cdots 1}\|) & \cdots & \sigma(\|\mathbf{t}_{21\cdots 1} - \mathbf{t}_{n_1\cdots n_d}\|) \\ \vdots & \vdots & \ddots & \vdots \\ \sigma(\|\mathbf{t}_{n_1\cdots n_d} - \mathbf{t}_{1\cdots 1}\|) & \sigma(\|\mathbf{t}_{n_1\cdots n_d} - \mathbf{t}_{21\cdots 1}\|) & \cdots & \sigma(\|\mathbf{t}_{n_1\cdots n_d} - \mathbf{t}_{n_1\cdots n_d}\|) \end{pmatrix}.$$

For a positive definite matrix $\mathbf{\Sigma}$ the root square $\mathbf{\Sigma}^{1/2}$ exists, where $\mathbf{\Sigma}^{1/2}$ is a matrix with the property $\mathbf{\Sigma}^{1/2}\mathbf{\Sigma}^{1/2} = \mathbf{\Sigma}$. For $i = 1, \ldots, (n_1\cdots n_d)$, let λ_i be the i-th eigenvalue of $\mathbf{\Sigma}$ associated with the normalized eigenvector $\mathbf{v}_i$, and let $\mathbf{D} := (\mathbf{v}_1 \cdots \mathbf{v}_{n_1\cdots n_d})$ and $\mathbf{\Lambda} := diag(\lambda_1 \cdots \lambda_{n_1\cdots n_d})$. Then we have a decomposition $\mathbf{\Sigma} = \mathbf{D\Lambda D}^\top$, where $\mathbf{D}$ is an orthogonal matrix. Hence, $\mathbf{\Sigma}^{1/2} = \mathbf{D\Lambda}^{1/2}\mathbf{D}^\top$, where $\mathbf{\Lambda}^{1/2} = diag(\sqrt{\lambda_1}\cdots\sqrt{\lambda_{n_1\cdots n_d}})$. Furthermore we define $\mathbf{\Sigma}^{-1/2}$ as a matrix with the property $\mathbf{\Sigma}^{-1/2} := (\mathbf{\Sigma}^{1/2})^{-1} = (\mathbf{\Sigma}^{-1})^{1/2}$. Therefore we have

$$\mathbf{\Sigma}^{-1/2} = ((\mathbf{D\Lambda D}^\top)^{1/2})^{-1} = (\mathbf{D\Lambda}^{1/2}\mathbf{D}^\top)^{-1} = \mathbf{D\Lambda}^{-1/2}\mathbf{D}^\top,$$

where $\mathbf{\Lambda}^{-1/2} = (\mathbf{\Lambda}^{1/2})^{-1} = diag(\frac{1}{\sqrt{\lambda_1}}\cdots\frac{1}{\sqrt{\lambda_{n_1\cdots n_d}}})$.

Upon transforming Model 3 using the linear mapping represented by the matrix $\mathbf{\Sigma}^{-1/2}$ we get a new model with independent and identically distributed errors

$$\mathbf{Y}^*_{n_1\cdots n_d} = \mathbf{X}^*_{n_1\cdots n_d}\beta + \mathcal{E}^*_{n_1\cdots n_d} \tag{4}$$

where $\mathbf{Y}^*_{n_1\cdots n_d} := \mathbf{\Sigma}^{-1/2}\mathbf{Y}_{n_1\cdots n_d}$, $\mathbf{X}^*_{n_1\cdots n_d} := \mathbf{\Sigma}^{-1/2}\mathbf{X}_{n_1\cdots n_d}$, and $\mathcal{E}^*_{n_1\cdots n_d} = \mathbf{\Sigma}^{-1/2}\mathcal{E}_{n_1\cdots n_d}$. For this transformed model the new vector of observations $\mathbf{Y}^*_{n_1\cdots n_d}$ is $n_1\cdots n_d$-dimensional normally distributed with $\mathbf{E}(\mathbf{Y}^*_{n_1\cdots n_d}) = \mathbf{X}^*_{n_1\cdots n_d}\beta$ and the variance-covariance matrix given by

$$Cov(\mathbf{Y}^*_{n_1\cdots n_d}) = Cov(\mathcal{E}_{n_1\cdots n_d}) = \sigma^2\mathbf{\Sigma}^{-1/2}\mathbf{\Sigma}\mathbf{\Sigma}^{-1/2} = \sigma^2\mathbf{I}.$$

This means that the components of $\mathbf{Y}^*_{n_1\cdots n_d}$ are uncorrelated so that under the normality condition they are independent and identically distributed with unit variance.

In the estimation of the parameter we consider only the transformed model. By the well-known Gauss-Marcov theorem (cf. Arnold [2]) the best unbiased estimator for the parameter β is given by the least squares estimator

$$\widehat{\beta} = (\mathbf{X}^{*\top}_{n_1\cdots n_d}\mathbf{X}^*_{n_1\cdots n_d})^{-1}\mathbf{X}^{*\top}_{n_1\cdots n_d}\mathbf{Y}^*_{n_1\cdots n_d} = (\mathbf{X}^{\top}_{n_1\cdots n_d}\mathbf{\Sigma}^{-1}\mathbf{X}_{n_1\cdots n_d})^{-1}\mathbf{X}^{\top}_{n_1\cdots n_d}\mathbf{\Sigma}^{-1}\mathbf{Y}_{n_1\cdots n_d}.$$

To test Hypothesis 2 we only need to consider the vector of the residuals of the transformed observations given by

$$\begin{aligned}\widehat{\mathbf{R}}^*_{n_1\cdots n_d} := (r^*_{j_1\cdots j_d})^{n_1,\ldots,n_d}_{j_1=1,\ldots,j_d=1} &= [\mathbf{I} - \mathbf{X}^*_{n_1\cdots n_d}(\mathbf{X}^{*\top}_{n_1\cdots n_d}\mathbf{X}^*_{n_1\cdots n_d})^{-1}\mathbf{X}^{*\top}_{n_1\cdots n_d}]\mathbf{Y}^*_{n_1\cdots n_d}\\ &= [\mathbf{I} - pr_{\mathbf{X}^*_{n_1\cdots n_d}}]\mathcal{E}^*_{n_1\cdots n_d},\end{aligned}$$

where $pr_{\mathbf{X}^*_{n_1\cdots n_d}}$ is the orthogonal projector onto the column space of $\mathbf{X}^*_{n_1\cdots n_d}$. This array of residuals is next embedded to a stochastic process having sample paths in the space $\mathcal{C}(\mathcal{A})$ by means of the set-indexed partial sums operator $\mathbf{V}_{n_1\cdots n_d} : \mathcal{R}^{n_1\times\cdots\times n_d} \to \mathcal{C}(\mathcal{A})$ defined by

$$\mathbf{V}_{n_1\cdots n_d}(\widehat{\mathbf{R}}^*_{n_1\cdots n_d})(B) = \sum_{j_1=1}^{n_1}\cdots\sum_{j_d=1}^{n_d}\sqrt{n_1\cdots n_d}\;\lambda^d(B\cap C_{j_1\cdots j_d})r^*_{j_1\cdots j_d},$$

where $\mathcal{A}$ is the family of compact subsets of $[\mathbf{a},\mathbf{b}]$, and $C_{j_1\cdots j_d} := \Pi^d_{k=1}[t_{nkj_k-1}, t_{nkj_k}]$ with $\lambda^d(C_{j_1\cdots j_d}) = 1/(n_1\cdots n_d)$, for $1 \le j_k \le n_k$, $k = 1,\ldots,d$. We refer interested reader to Pyke [13], Alexander and Pyke [1], and recently to Somayasa [17] for more detailed discussion regarding this process.

We define reasonable statistics for testing (2) which are Kolmogorov-Smirnov (KS) and Cramér-von Misses (CvM) functionals of the process. That are

$$\begin{aligned} KS_{n_1\cdots n_d;\mathcal{A}} &:= \sup_{B\in\mathcal{A}}|\frac{1}{\sigma}\mathbf{V}_{n_1\cdots n_d}(\widehat{\mathbf{R}}^*_{n_1\cdots n_d})(B)|\\ CvM_{n_1\cdots n_d;\mathcal{A}} &:= \frac{1}{n_1\cdots n_d}\sum_{B\in\mathcal{A}}(\frac{1}{\sigma}\mathbf{V}_{n_1\cdots n_d}(\widehat{\mathbf{R}}^*_{n_1\cdots n_d})(B))^2.\end{aligned}$$

Asymptotic test using the KS test will reject H_0 at a level of significance α if and only if $KS_{n_1\cdots n_d;\mathcal{A}} \ge q_{1-\alpha}$, where $q_{1-\alpha}$ is the $(1-\alpha)$ quantile of the limit distribution of $KS_{n_1\cdots n_d;\mathcal{A}}$. Similarly, using the CvM test, H_0 will be rejected at a level of significance α if and only if $CvM_{n_1\cdots n_d;\mathcal{A}} \ge c_{1-\alpha}$, where $c_{1-\alpha}$ is the $(1-\alpha)$ quantile of the limit distribution of $CvM_{n_1\cdots n_d;\mathcal{A}}$.

2.2. Extension to multivariate case

In the multivariate case we observe a vector of spatial processes $\{\mathbf{Y}(\mathbf{t}) := (Y_i(\mathbf{t}))^p_{i=1} : \mathbf{t}\in[\mathbf{a},\mathbf{b}]\}$ on the experimental design $\Xi_{n_1\cdots n_d}$ to test the hypothesis defined as follows

$$H_0 : \mathbf{g} = (g_i)^p_{i=1} \in \prod_{i=1}^p \mathbf{W}_i \;\; against \;\; H_1 : \mathbf{g} = (g_i)^p_{i=1} \notin \prod_{i=1}^p \mathbf{W}_i, \tag{5}$$

where for every $\mathbf{t}\in[\mathbf{a},\mathbf{b}]$, $\mathbf{Y}(\mathbf{t})$ allows a decomposition $\mathbf{Y}(\mathbf{t}) = \mathbf{g}(\mathbf{t}) + \zeta(\mathbf{t})$ for some vector of random errors $\zeta := (\zeta_1,\ldots,\zeta_p)^\top$ which follows a p-variate normal distribution with mean vector $\mathbf{0}$ and the variance-covariance matrix defined below. In this case the existence of correlation among the components of $\mathbf{Y}$ and among the observations is inherent. That is for any $\mathbf{t}_\ell, \mathbf{t}_k \in [\mathbf{a},\mathbf{b}]$,

$Cov(\mathbf{Y}(\mathbf{t}_\ell),\mathbf{Y}(\mathbf{t}_k)) = \mathbf{\Sigma}(\mathbf{t}_\ell - \mathbf{t}_k)$, where $\mathbf{\Sigma}(\mathbf{t}_\ell - \mathbf{t}_k)$ is a $p \times p$-dimensional matrix of functions of the difference $(\mathbf{t}_\ell - \mathbf{t}_k)$ taking the form

$$\mathbf{\Sigma}(\mathbf{t}_\ell - \mathbf{t}_k) = \begin{pmatrix} \sigma_{11}(\mathbf{t}_\ell - \mathbf{t}_k) & \sigma_{12}(\mathbf{t}_\ell - \mathbf{t}_k) & \cdots & \sigma_{1p}(\mathbf{t}_\ell - \mathbf{t}_k) \\ \sigma_{21}(\mathbf{t}_\ell - \mathbf{t}_k) & \sigma_{22}(\mathbf{t}_\ell - \mathbf{t}_k) & \cdots & \sigma_{2p}(\mathbf{t}_\ell - \mathbf{t}_k) \\ \vdots & \vdots & \ddots & \vdots \\ \sigma_{p1}(\mathbf{t}_\ell - \mathbf{t}_k) & \sigma_{p2}(\mathbf{t}_\ell - \mathbf{t}_k) & \cdots & \sigma_{pp}(\mathbf{t}_\ell - \mathbf{t}_k) \end{pmatrix}.$$

To simplify the notation and also for computational reason, we assume that $\mathbf{W}_1 = \cdots = \mathbf{W}_p$. Hence we get a more simple model

$$\mathbf{Y}(\ell/n, k/n) = \mathbf{g}(\ell/n, k/n) + \zeta(\ell/n, k/n),\ 1 \le \ell, k \le n.$$

For a fixed $n \ge 1$, let $\mathbf{Y}_{nnp} := (vec^\top(Y_1^\top(\Xi_n)), vec^\top(Y_2^\top(\Xi_n)), \ldots, vec^\top(Y_p^\top(\Xi_n)))^\top$ be pn^2-dimensional vector of random observations, where $Y_i(\Xi_n) := (Y_i(\ell/n, k/n))_{k=1,\ell=1}^{n,n}$, and "vec" is the well-known vec operator (cf. Harville [9]). Then the model under H_0 can be represented as

$$\mathbf{Y}_{nnp} = \mathbf{A}_{pn^2 \times p^2}\beta + \zeta_{nnp} \tag{6}$$

where $\beta := (\beta_1^\top, \ldots, \beta_p^\top)^\top$ is the p^2-dimensional vector of parameters, with $\beta_i := (\beta_{1i}, \cdots, \beta_{pi})^\top$, $\mathbf{A}_{pn^2 \times p^2} := diag(\mathbf{X}_{nn}, \cdots, \mathbf{X}_{nn})$ is the $pn^2 \times p^2$-dimensional block matrix, and ζ_{nnp} is the associated pn^2-dimensional normally distributed vector of random errors with $\mathbf{E}(\zeta_{nnp}) = \mathbf{0}$ and $Cov(\zeta_{nnp}) = \mathbf{\Sigma}_{nnp}$, where $\mathbf{\Sigma}_{nnp}$ is the $pn^2 \times pn^2$-dimensional unknown matrix. For further simplification of the problem we assume the covariance matrix $\mathbf{\Sigma}_{nnp}$ has a structure $\mathbf{\Sigma}_{nnp} = \mathbf{\Sigma}_p \otimes \Psi_{nn}$, where $\otimes$ stands for the well-known Kronecker product defined e.g. in Harville [9] given by

$$\mathbf{\Sigma}_p \otimes \Psi_{nn} = \begin{pmatrix} \sigma_{11}\Psi_{nn} & \sigma_{12}\Psi_{nn} & \cdots & \sigma_{1p}\Psi_{nn} \\ \sigma_{21}\Psi_{nn} & \sigma_{22}\Psi_{nn} & \cdots & \sigma_{2p}\Psi_{nn} \\ \vdots & \vdots & \vdots & \vdots \\ \sigma_{p1}\Psi_{nn} & \sigma_{p2}\Psi_{nn} & \cdots & \sigma_{pp}\Psi_{nn} \end{pmatrix},$$

σ_{ij} represents the covariance between Y_i and Y_j, and Ψ_{nn} is the $n^2 \times n^2$-dimensional matrix of covariance functions taking the form

$$\Psi_{nn} = \begin{pmatrix} \psi(\mathbf{t}_{n11} - \mathbf{t}_{n11}) & \cdots & \psi(\mathbf{t}_{n11} - \mathbf{t}_{nn1}) & \cdots & \psi(\mathbf{t}_{n11} - \mathbf{t}_{n1n}) & \cdots & \psi(\mathbf{t}_{n11} - \mathbf{t}_{nnn}) \\ \vdots & & \vdots & & \vdots & & \vdots \\ \psi(\mathbf{t}_{n1k} - \mathbf{t}_{n11}) & \cdots & \psi(\mathbf{t}_{n1k} - \mathbf{t}_{nn1}) & \cdots & \psi(\mathbf{t}_{n1k} - \mathbf{t}_{n1k}) & \cdots & \psi(\mathbf{t}_{n1k} - \mathbf{t}_{nnn}) \\ \vdots & & \vdots & & \vdots & & \vdots \\ \psi(\mathbf{t}_{nnn} - \mathbf{t}_{n11}) & \cdots & \psi(\mathbf{t}_{nnn} - \mathbf{t}_{nn1}) & \cdots & \psi(\mathbf{t}_{nnn} - \mathbf{t}_{n1n}) & \cdots & \psi(\mathbf{t}_{nnn} - \mathbf{t}_{nnn}) \end{pmatrix},$$

where for $1 \le \ell, k \le n$, $\mathbf{t}_{n\ell k} := (\ell/n, k/n)$. Note that in the case of spatial isotropy one has

$$\Psi_{nn} = (\psi(\|\mathbf{t}_{n\ell k} - \mathbf{t}_{n\ell' k'}\|))_{\ell=1,k=1,\ell'=1,k'=1}^{n,n,n,n} \in \mathcal{R}^{n^2 \times n^2}.$$

As in the univariate case, under the assumption that $\mathbf{\Sigma}_p \otimes \Psi_{nn}$ is positive definite, upon pre-multiplying Model 6 by $(\mathbf{\Sigma}_p \otimes \Psi_{nn})^{-1/2}$, we get the transformed model as

$$(\mathbf{\Sigma}_p \otimes \Psi_{nn})^{-1/2}\mathbf{Y}_{nnp} = (\mathbf{\Sigma}_p \otimes \Psi_{nn})^{-1/2}\mathbf{A}_{pn^2 \times p^2}\beta + (\mathbf{\Sigma}_p \otimes \Psi_{nn})^{-1/2}\zeta_{nnp}. \tag{7}$$

By using such transformation, the new model $\mathbf{Y}^*_{nnp} = \mathbf{A}^*_{pn^2 \times p^2}\beta + \zeta^*_{nnp}$ consists of the vector of random errors ζ^*_{nnp} owning the variance-covariance matrix $\mathbf{I}_{pn^2 \times pn^2}$. Hence, by the Gauss-Markov theorem the least squares estimator of β, denoted by $\widehat{\beta}$ will become the BLUE of β which is given by

$$\widehat{\beta} = (\mathbf{A}^\top_{pn^2 \times p^2}(\mathbf{\Sigma}_p \otimes \Psi_{nn})^{-1}\mathbf{A}_{pn^2 \times p^2})^{-1}\mathbf{A}^\top_{pn^2 \times p^2}(\mathbf{\Sigma}_p \otimes \Psi_{nn})^{-1}\mathbf{Y}_{nnp}.$$

Similarly, we get the corresponding vector of the least squares residuals as

$$\widehat{\mathbf{R}}^*_{nnp} := [\mathbf{I}_{pn^2 \times pn^2} - \mathbf{A}^*_{pn^2 \times p^2}(\mathbf{A}^{*\top}_{pn^2 \times p^2}\mathbf{A}^*_{pn^2 \times p^2})^{-1}\mathbf{A}^{*\top}_{pn^2 \times p^2}]\zeta^*_{nnp}.$$

The transformation defined above also implies that the array $\{\zeta^*(\ell/n, k/n) = (\zeta^*_i(\ell/n, k/n))^p_{i=1} : 1 \leq \ell, k \leq n\}$ consists of independent and identically normally distributed random vectors with $\mathbf{E}(\zeta^*(\ell/n, k/n)) = \mathbf{0}$ and $Cov(\zeta^*(\ell/n, k/n)) = \mathbf{I}_{p \times p}$.

The p-dimensional set-indexed partial sums operator is denoted by $\mathbf{V}_{nnp}$ defined on the space $\prod_{i=1}^p \mathcal{R}^{n^2}$ taking values in the space $\prod_{i=1}^p \mathcal{C}(\mathcal{A})$ given by

$$\mathbf{V}_{nnp}(\widehat{\mathbf{R}}^*_{nnp})(B) = \sum_{k=1}^n \cdots \sum_{\ell=1}^n n\lambda^2(B \cap C_{\ell k})\mathbf{r}^*_{\ell k}, \tag{8}$$

where $C_{\ell k} := ((\ell - 1)/n, \ell/n] \times [(k-1)/n, k/n]$, and $\mathcal{A}$ is the family of compact subsets of $[0,1] \times [0,1]$. Interested reader is referred to Somayasa [18] for the definition of (8). Note that $\mathbf{r}^*_{\ell k}$ is the vector of the residual in the point $(\ell/n, k/n)$ associated to the transformed model. The Kolmogorov-Smirnov (KS) and Cramér-von Misses (CvM) functionals of the p-dimensional process are defined as follows

$$KS_{nnp;\mathcal{A}} := \sup_{B \in \mathcal{A}} \|\mathbf{V}_{nnp}(\widehat{\mathbf{R}}^*_{nnp})(B)\| \text{ and } CvM_{nnp;\mathcal{A}} := \frac{1}{n^2}\sum_{B \in \mathcal{A}} \|\mathbf{V}_{nnp}(\widehat{\mathbf{R}}^*_{nnp})(B)\|^2.$$

Similarly, asymptotic test using the KS test will reject H_0 at a level of significance α if and only if $KS_{nnp;\mathcal{A}} \geq t_{1-\alpha}$, where $t_{1-\alpha}$ is the $(1-\alpha)$ quantile of the limit distribution of $KS_{nnp;\mathcal{A}}$. Using the CvM test, H_0 will be rejected at a level of significance α if and only if $CvM_{nnp;\mathcal{A}} \geq w_{1-\alpha}$, where $w_{1-\alpha}$ is the $(1-\alpha)$ quantile of the limit distribution of $CvM_{nnp;\mathcal{A}}$.

3. Limiting behavior of the tests

In this section we investigate the limit process of the sequence of set-indexed least squares residual partial sums process defined in the preceding section. Although in one hand the distribution model of the sample is known to be normal, but on the other hand we insists on the derivation of the asymptotic behavior of the KS and CvM tests.

Firstly let us consider the univariate case represented by Model 4 and without loss of the generality we consider the unit rectangle $[0,1] \times [0,1]$ and Ξ_n as the experimental region and design, respectively. Let $\mathbf{W}^*_{nn} := [f^*_1(\Xi_{nn}), \ldots, f^*_p(\Xi_{nn})]$, where $f^*_i(\Xi_{nn}) := \mathbf{\Sigma}^{-1/2}f_i(\Xi_{nn})$ is a vector in $\mathcal{R}^{n^2}$, $i = 1, \ldots, p$ which are assumed to be orthonormal. Then we have

$$\widehat{\mathbf{R}}^*_{nn} = \mathcal{E}^*_{nn} - pr_{\mathbf{W}^*_{nn}}\mathcal{E}^*_{nn} = \mathcal{E}^*_{nn} - \sum_{i=1}^p \langle f^*_i(\Xi_{nn}), \mathcal{E}^*_{nn}\rangle_{\mathcal{R}^{n^2}} f^*_i(\Xi_{nn}).$$

Hence by the linearity of $\mathbf{V}_{nn}$ and by referring to Lemma 5.1 in Somayasa [17] we further get

$$\begin{aligned}\mathbf{V}_{nn}(\widehat{\mathbf{R}}^*_{nn})(B) &= \mathbf{V}_{nn}(\mathcal{E}^*_{nn})(B) - \sum_{i=1}^p \langle \mathbf{V}_{nn}(f^*_i(\Xi_{nn})), \mathbf{V}_{nn}(\mathcal{E}_{nn})\rangle_{\mathcal{H}_Z} \mathbf{V}_{nn}(f^*_i(\Xi_{nn}))(B) \\ &= \mathbf{V}_{nn}(\mathcal{E}^*_{nn})(B) - (pr_{\mathbf{V}_{nn}(\mathbf{W}^*_{nn})}\mathbf{V}_{nn}(\mathcal{E}^*_{nn}))(B), \ \forall B \in \mathcal{A},\end{aligned}$$

where $\mathbf{V}_{nn}(\mathbf{W}^*_{nn}) := [\mathbf{V}_{nn}(f_1^*(\Xi_{nn})), \ldots, \mathbf{V}_{nn}(f_p^*(\Xi_{nn}))] \subseteq \mathcal{H}_Z \subset \mathcal{C}(\mathcal{A})$. For $i = 1, \ldots, p$, let $s_i^{(n)}$ be a step function on $[0,1] \times [0,1]$ associated to $f_i^*(\Xi_{nn})$, defined by

$$s_i^{(n)}(t,s) := \sum_{\ell=1}^{n}\sum_{k=1}^{n} f_i^*(\ell/n, k/n)\mathbf{1}_{C_{\ell k}}(t,s),$$

then we have $h_{s_i^{(n)}}(B) := \int_B s_i^{(n)}(t,s)\lambda^2(dt,ds) = \frac{1}{n}\mathbf{V}_{nn}(f_i^*(\Xi_{nn}))$. Hence the set $\{h_{s_1^{(n)}}, \ldots, h_{s_p^{(n)}}\}$ also builds a basis for the space $\mathbf{V}_{nn}(\mathbf{W}^*_{nn})$. Now we are ready to state our main result which can be proved in the similar way as in the result of Somayasa [17].

Theorem 3.1 Let $f_1^*, \ldots, f_p^*$ be orthonormal as functions in $L_2(\lambda^2, [0,1] \times [0,1])$, where f_i^* is obtained in such a way that $\|s_i^{(n)} - f_i^*\|_\infty$ converges to zero. If $f_1, \ldots, f_p$ are continuous and have bounded variation on $[0,1] \times [0,1]$ and $\boldsymbol{\Sigma}$ is positive definite, then

$$\frac{1}{\sigma}\mathbf{V}_{nn}(\widehat{\mathbf{R}}^*_{nn})(\cdot) \Rightarrow Z_{\mathbf{f}^*}(\cdot) := Z(\cdot) - \sum_{i=1}^{p}\left(\int_{[0,1]\times[0,1]}^{(R)} f_i^*(t,s)dZ(t,s)\right)h_{f_i^*}(\cdot),$$

where $Z(\cdot)$ is the centered Gaussian process indexed by $\mathcal{A}$, see [1, 13] for the definition of $Z(\cdot)$.

The following result is an immediate consequence of the well-known continuous mapping theorem, see e.g. Theorem 5.1 in Billingsley [4].

Corollary 3.2 Under the condition of Theorem 3.1 it holds

$$KS_{nn;\mathcal{A}} \Rightarrow \sup_{B\in\mathcal{A}} |Z_{\mathbf{f}^*}(B)| \text{ and } CvM_{nn} \Rightarrow \int_{[0,1]\times[0,1]} Z^2_{\mathbf{f}^*}(B)dB.$$

Corollary 3.2 suggests us to approximate the quantiles of $KS_{nn;\mathcal{A}}$ and CvM_{nn} by those of the distributions of the statistics $\sup_{B\in\mathcal{A}}|Z_{\mathbf{f}^*}(B)|$ and $\int_{[0,1]\times[0,1]} Z^2_{\mathbf{f}^*}(B)dB$ which can be directly derived from that of the set-indexed Brownian sheet. However in this work the finite sample quantiles of the statistics $KS_{nn;\mathcal{A}}$ and CvM_{nn} will be approximated by simulation using computer.

Next we extend our consideration to the multivariate model. The p-dimensional set-indexed partial sums operator on $\prod_{i=1}^{p} \mathcal{R}^{n^2}$ is defined component wise as in Equation 8. Hence the limit processes can also directly derived by using the convergence component wise.

Theorem 3.3 Let $f_1^*, \ldots, f_p^*$ be orthonormal as functions in $L_2(\lambda^2, [0,1] \times [0,1])$, where f_i^* is obtained in such a way that $\|s_i^{(n)} - f_i^*\|_\infty$ converges to zero. If $f_1, \ldots, f_p$ are continuous and have bounded variation on $[0,1] \times [0,1]$ and $\boldsymbol{\Sigma}_p \otimes \boldsymbol{\Psi}_{nn}$ is positive definite, then

$$\mathbf{V}_{nnp}(\widehat{\mathbf{R}}^*_{nnp})(\cdot) \Rightarrow \mathbf{Z}^{(p)}_{\mathbf{f}^*}(\cdot) := \mathbf{Z}^{(p)}(\cdot) - \left(\sum_{j=1}^{p}\left(\int_{[0,1]\times[0,1]}^{(R)} f_j^*(t,s)dZ^{(i)}(t,s)\right)h_{f_j^*}(\cdot)\right)_{i=1}^{p},$$

where $\mathbf{Z}^{(p)}(\cdot) := (Z^{(1)}, \cdots, Z^{(p)})^\top$ is the p-dimensional centered Gaussian process indexed by $\mathcal{A}$, defined e.g. in Somayasa and Wibawa [18]. We notice that the convergence $\mathbf{V}_{nnp}(\zeta^*_{nnp})(\cdot) \Rightarrow \mathbf{Z}^{(p)}(\cdot)$ can be shown by extending the uniform central limit theorem studied in [1] and [13] to the vectorial analog.

As a direct consequence of the continuous mapping theorem, we get the following corollary.

Journal of Physics: Conference Series **725** (2016) 012003 doi:10.1088/1742-6596/725/1/012003

Corollary 3.4 Under the condition of Theorem 3.1 it holds

$$KS_{nnp;\mathcal{A}} := \sup_{B\in\mathcal{A}} \|\mathbf{V}_{nnp}(\widehat{\mathbf{R}}^*_{nnp})(B)\| \Rightarrow \sup_{B\in\mathcal{A}} \|\mathbf{Z}^{(p)}_{\mathbf{f}^*}(B)\|$$

$$CvM_{nnp;\mathcal{A}} := \frac{1}{n^2}\sum_{B\in\mathcal{A}} \left\|\mathbf{V}_{nnp}(\widehat{\mathbf{R}}^*_{nnp})(B)\right\|^2 \Rightarrow \int_{[0,1]\times[0,1]} \|\mathbf{Z}_{\mathbf{f}^*}(B)\|^2 \, dB.$$

Corollary 3.4 give us important way in approximating the sampling distributions of the multivariate KS and CvM statistics. In the practice the tests are realized by computing the quantiles of the limiting statistics presented in the right-hand side of Corollary 3.4. Unfortunately they are mathematically not tractable. By this reason in this work we develop simulation for approximating the quantiles of the statistics.

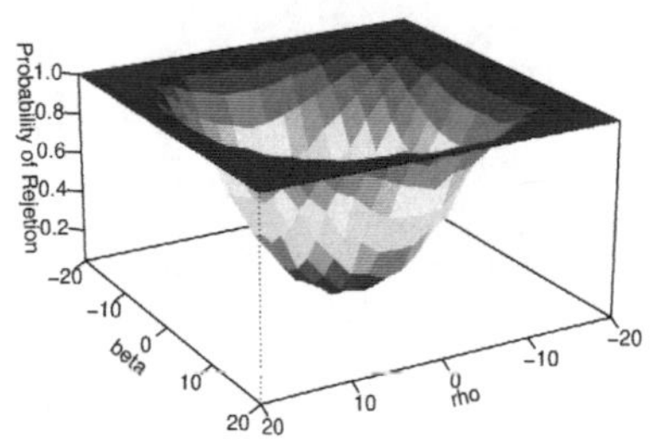

(a) Power function of KS test

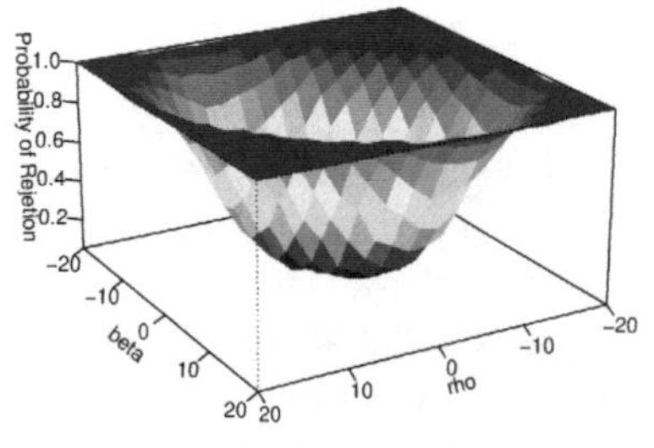

(a) Power function of CvM test

Figure 1. The empirical power functions of the KS and CvM tests for univariate constant model.

4. Simulation study

In this section we study the finite sample properties of both tests by using Monte Carlo simulation using software package R. However by technical and computational reasons we only consider univariate and bivariate constant models, although theoretically we have more options for our choice such as any polynomial of finite degree.

4.1. Simulation 1

We considering under H_0 a univariate constant model $Y(t,s) = 5 + \varepsilon(t,s)$, while under H_1 we assume a first order model $Y(t,s) = 5 + \rho t - \beta s + \varepsilon(t,s)$, for $(t,s) \in [0,1]\times[0,1]$, where $\varepsilon(t,s)$ is assumed to be normally distributed and isotropic having a covariance function

$$Cov(\varepsilon(t,s), \varepsilon(t',s')) = \sigma^2 \exp\{-\|(t,s)-(t',s')\|\}, \ (t,s),(t',s') \in [0,1]\times[0,1].$$

Thereby σ^2 is an unknown parameter which is estimated by a consistent estimator defined in Arnold [2]. The samples are generated from a localized model $Y(\ell/n, k/n) = (5 + \rho\ell/n + \beta k/n)/n + \varepsilon(\ell/n, k/n)$, for $1 \le \ell, k \le n$, under $n \times n$ regular lattice Ξ_n. It is clear that the model is from H_0 if and only if ρ and β are simultaneously zero, otherwise the model is from H_1. This means for a predetermined α, the power functions must attain α when both ρ and β are zero. Conversely, when either α or β are not zero, the power functions must take values larger than α.

The three dimensional empirical power functions of the KS and CvM tests are presented in Figure 1. The graphs are generated by drawing a surface trough the points $(\rho, \beta, P(\rho, \beta))$ using three dimensional perspective plot of R, where for a fixed (ρ, β), $P(\rho, \beta)$ is the relative number of the values of $KS_{nn;\mathcal{A}}$ (respectively $CvM_{nn;\mathcal{A}}$) exceeding the 0.95 quantile of $KS_{nn;\mathcal{A}}$ (respectively $CvM_{nn;\mathcal{A}}$). Figure 1 shows that both tests attain the level of significance $\alpha = 0.05$ as ρ and β are fixed to zero. When either ρ or β moves away from zero, the probability of rejection of H_0 gets larger. In general, for the constant model the KS test shows larger power than the CvM test.

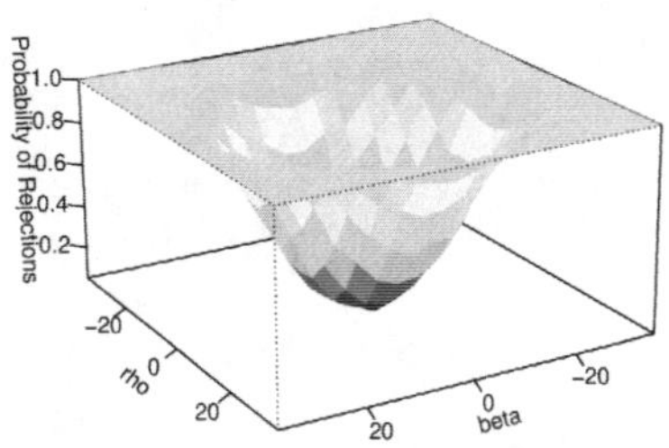

(a). Power function of KS test

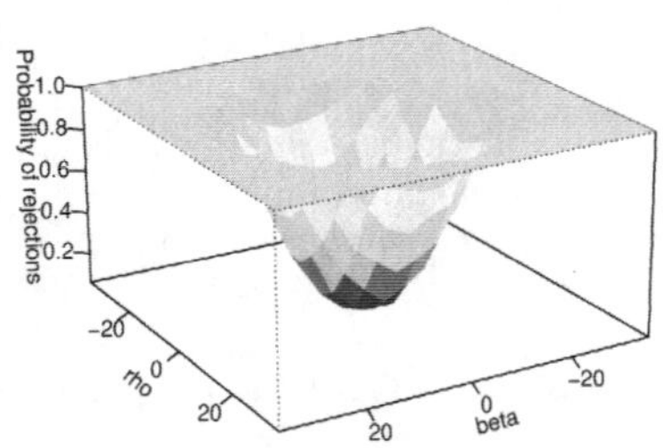

(b). Power function of CvM test

Figure 2. The empirical power functions of the KS and CvM tests for bivariate constant model.

4.2. Simulation 2

In the second case we simulate a multivariate constant model with two responses. The sample from localized bivariate model $\mathbf{Y}(\ell/n, k/n) = (5+\rho k/n+\rho\ell/n, 3+\beta k/n+\beta\ell/n)/n+\zeta(\ell/n, k/n)$. To simplify the computation, we assume that $\Psi_{nn} = \mathbf{I}_{n^2\times n^2}$ and the random vector $\zeta = (\zeta_1, \zeta_2)^\top$ is generated form a bivariate normal distribution with mean zero and variance-covariance matrix

$$\mathbf{\Sigma} := \begin{pmatrix} +6.26 & -0.50 \\ -0.50 & +6.26 \end{pmatrix}.$$

However $\mathbf{\Sigma}$ is assumed to be unknown and for that it is estimated by a consistent estimator proposed in Arnold [3].

Figure 2 exhibits the three-dimensional graphics of the empirical power function of the tests, where the first graph (a) represents the power function of the KS test and the second (b) represents that of the CvM test. Both power functions seem to achieve the pre signed level of significance $\alpha = 0.05$. The power of the tests get larger as the the model moves away from H_0. Among the two tests the CvM test tends to have larger power.

5. Conclusion

We established asymptotic test procedures based on the Kolmogorov-Smirnov and Cramér-von Mises functionals of the partial sums processes of the least squares residuals. We obtain the limit processes as function of the set-indexed Brownian sheet for univariate case, whereas for multivariate case the limit process is a functional of the p-dimensional Brownian sheet indexed by sets. Observing constant model the finite sample size behavior of the tests are that in the univariate case the KS test tend to have a larger power than that of CvM test. Conversely in multivariate case the CvM test is slightly more powerful then the KS test. By this reason in the application we suggest the user to use the KS test rather than the CvM test for univariate constant model. However for bivariate constant model we propose to adopt the CvM test.

Acknowledgement

This work was supported in parts by SAME program 2015. The authors wish to thank the Institute for Stochastic Karlsruhe Institute of Technology (KIT) for hospitality.

References

[1] Alexander K S and Pyke R 1986 *The Annals of Probability* **14** 582–97.
[2] Arnold S F 1981 *The Theory of Linear Models and Multivarite Analysis* (New York: John Wiley & Sons, Inc.)
[3] Arnold S F 1984 *Journal of the American Statistical Association* 75 **(372)** 890–94.
[4] Billingsley P 1968 *Convergence of Probability Measures* (New York: John Wiley & Sons, Inc.)
[5] Bischoff W 2002 *Theory of Stochastic Processes* **2** 23–8
[6] Bischoff W and Somayasa W 2009 *J. Multivariate Analysis* **100** 2167–77
[7] Christensen R 1991 *Linear Models for Multivariate, Time Series, and Spatial Data* (New York:Springer)
[8] Cressie Noel A C 1991, *Statistics for Spatial Data* (New York:John Wiley & Sons)
[9] Harville D A 1997 *Matrix Algebra from a Statistician's Perspective* (New York: Springer-Verlag)
[10] Graybill F A 2000 *Theory and Application of Linear Model* (London:Duxbury press)
[11] Johnson R A and Wichern D W 2007 *Applied Multivariate Statistical Analysis* (New York: Prentice-Hall)
[12] MacNeill I B and Jandhyalla V K 1993 *Multivariate Environmental Statistics* ed G P Patil and C R Rao (Amsterdam: Elsevier Science Publishers B.V.) 298–306.
[13] Pyke R 1983 *Probability, Statistics and Analysis* ed J F C Kingman and G E H Reuter (London: Math. Soc. Lect. Notes Series) **79** 219–40.
[14] Schabenberger S and Gotway C A 2005 *Statistical Methods for Spatial Data Analysis* (London:Chapman)
[15] Seber G A F and Lee A J 2003 *Linear Regression Analysis* (2nd ed.) (New York: John Wiley & Sons)
[16] Somayasa W 2014 *J. Indones. Math.Soc.* 20**(2)** 77–94.
[17] Somayasa W, Ruslan, Cahyono E and Engkoimani L O 2015 *Fareast J. Math. Sciences* 96**(8)** 933–66
[18] Somayasa W, and Wibawa G N A 2015 *Fareast J. Math. Sciences* 98**(5)** 613–39
[19] Stute W 1997 *The Annals of Statistics* 25**(2)** 613–41
[20] Stute W, Mantaiga W G and Quindimil M P 1997 *J. Amer. Statistical Association* 93**(441)** 141–9
[21] Xie L and MacNeill I B 2006 *South African Statist. J.* **4** 33–3
[22] Zellner A 1962 *J. Amer. Statistical Association* 57 **(298)** 348–68

Journal of Physics: Conference Series **725** (2016) 012004 doi:10.1088/1742-6596/725/1/012004

A Novel Evaluation algorithm of Urban Comprehensive Carrying Capacity Index System in Northwest China:Lanzhou city as a example

Jing Li[1,2] *, Song Lei[1], Ma Ling[1]

[1]School of Economics and Management, Lanzhou City University, Lanzhou 730070, Gansu China

[2]School of Urban Economics and Tourism culture, Lanzhou City University, Lanzhou 730070, Gansu China;

E-mail:jinglilzu@163.com

Abstract. With the rapid growth of the urban population and quick expansion of urban dimensions, the affecting urban development constraints of urban carrying capacity has increasingly been attracting people's attention. However, the effective synthesization of urban resources carrying capacity, urban environmental carrying capacity, economic carrying capacity, society carrying capacity and traffic carrying capacity hardly acted as an urban comprehensive carrying capacity evaluation system. In this paper, the novel comprehensive index system of urban resources carrying capacity, urban environmental carrying capacity, economic carrying capacity, society carrying capacity and traffic carrying capacity was used for analyzing and evaluating urban comprehensive carrying capacity. Firstly, the standardization of evaluation index was constructed in the urban comprehensive carrying capacity index system. In addition, the past decade data from 2002 to 2011 was analyzed and evaluated in the weighted index system. Finally, the influence mechanism of various carrying capacity factors on urban comprehensive carrying capacity was investigated in Lanzhou. Experimental results can be summarized that the variation tendency of urban comprehensive carrying capacity was the annual increase of levels and the promotion countermeasure was put forward with congenital restricting factors and acquired restricting factors.

1. Introduction

City is the concentration area of population distribution, resources consumption and environmental pollution. Moreover, it is the most important carrier of social life of human and economic development. In the 21st century, China has entered into a phase of rapid urbanization which brings rapid growth of urban population and the rapid expansion of city scale. The urban economic and social structure transforms rapidly, public demands experiences comprehensive and rapid growth, the constraint of resources and environment further tightens, business costs keeps rising. The city is facing increasingly serious ecological environment problems with fragility of city safety and disaster prevention. All of these has seriously affected the sustainable utilization of natural resources and the sustainable development of urban economy, society, ecology and culture[1].Lanzhou is the capital city of Gansu province. In the 12th five-year plan, urbanization rate will reach 80% in 2015 which the net urbanization rate, rate is 5% compared with the 11th five-year planand the city district population will reach 4.2 million, a net increase of 860000 compared with that of 2010. Under such a heavy population pressure,

Content from this work may be used under the terms of the Creative Commons Attribution 3.0 licence. Any further distribution of this work must maintain attribution to the author(s) and the title of the work, journal citation and DOI.
Published under licence by IOP Publishing Ltd

the city sustainable development can coordinate with the urban comprehensive carrying capacity not only is related with its own development trends, but also is related with the successful sustainable development of surrounding areas. Therefore it is important to study the changes of Lanzhou's urban comprehensive carrying capacity, and forecast its future development.

The concept of "carrying-capacity" firstly appeared in the field of ecology. It is used to measure the maximum number of individuals within a species in specific area under certain environment conditions. After entering into the 21st century, the concept of "carrying-capacity" was introduced to the urban system. In January 2005, the concept of urban comprehensive carrying capacity was clearly put forward in the *"notice about strengthening the overall urban planning and revision and the work of examination and approval"* issued by the ministry of construction. A growing number of scholars at home and abroad express their opinions, and conduct research on this problem at multiple levels. Some scholars take the urban land resources, water resources, environment and resources, mineral resources, cultural resources, and other single factors as the breakthrough point to carry on empirical analysis [2-9].Such kind of research is mainly based on the principle of minimum limiting factor, stressing that the most scarce resources have a decisive impact on urban comprehensive carrying capacity. But it should be clear that for the city which is development system, the influence of single factor on urban comprehensive carrying capacity is significantly weakened due to the flow and complementation of elements. But the synergy effect of multiple factor is significantly enhancing. At the same time, the abundance of natural resources and the city's economic development level is not completely in positive correlation [1]. Although a lot of places in western China, such as Yunnan and Guizhou, these places have beautiful mountains and rivers and abundant natural resources, but their economic development level is much lower than other advanced cities. And the economic development level of the cities like Tokyo in Japan and Shenzhen in China is extremely high, although they have scarce natural resources. According to it, many scholars try to build the evaluation system from the perspective of multiple factors and carry out empirical analysis and quantitative analysis on urban comprehensive carrying capacity[10-12].This kind of research is mainly based on the compensation effect or comprehensive effect of resources carrying capacity[6], and study the influencing degree of various factors on urban comprehensive carrying capacity through factor weight, and further reflects the relationship and compensation among various factor. In other words, when a factor is restricted, it can be compensated through strengthening it. Take the city's natural composite system as an example. Tree-planting, a forestation, and water conservation not only can strengthen the protection of the environment, but also further improve the capacity of water resources supply and production capacity of land resources within certain area. Some scholars carry out a series of beneficial exploration on the study of the carrying-capacity of urban agglomeration on the basis of the above theory and method , discusses the relationship between regional economic development and environment coordination [13-14].In the study of the calculation method and model of carrying-capacity, the scholars adopts the prediction method of ecological footprint, logistic model, system dynamics method, the GM (1, 1) mode in evaluation [15].But in this type of research, clear difference can be seen from the evaluation results concluded from different scholars and it is worth noting that the evaluation result has large deviation with the reality [16].

It should be pointed out that, although the study of urban comprehensive carrying capacity has achieved certain progress, but the theoretical framework and method system which can be widely accepted and get better reflect in the process of practical application has not been set up.This paper argues that there are several problems need to pay attention to in future research: (1) the structural problems of the carrying-capacity of cities in different area are different. And the main restricting factors are different, thus different index systems should be built for evaluation. It is obvious that the level of the urban comprehensive carrying capacity of cities in eastern region has significant differences with that of the cities in the western region.(2) the structural problems of the carrying-capacity of cities in different area are different, thus different index systems should be built for evaluation. The paper believes that the scale here should be divided on the basis of urban population rather than on the basis of urban administrative levels. There are many eastern counties have a population scale over millions, while the population of many western cities only are hundreds of thousands of. Obviously, there are no much comparison to use urban administrative levels to carry out analysis of the level of carrying-capacity.(3) The structural problem of urban comprehensive carrying capacity at different development stages is not the same, thus different index systems should be built for evaluation. The city focusing on development speed is different from the city focusing on development quality in the aspect of main constraint factors, thus their methods of improving urban comprehensive carrying capacity are different.(4)The urban comprehensive carrying capacity of the same city will be different if it stays at different functional zones. The evaluation should be based on classification quantification.(5) The urban public service quality and management level of urban government has significant impact on urban comprehensive carrying capacity, but they have totally different influences on cities with different locations, cities with different scales, cities at different development stage. Besides, the determination of weight is also different.

In this paper, the novel comprehensive index system of urban resources carrying capacity, urban environmental carrying capacity, economic carrying capacity, society carrying capacity and traffic carrying capacity was used for analyzing and evaluating urban comprehensive carrying capacity. Firstly, the standardization of evaluation index was constructed in the urban comprehensive carrying capacity index system. In addition, the past decade data from 2002 to 2011 was analyzed and evaluated in the weighted index system. Finally, the influence mechanism of various carrying capacity factors on urban comprehensive carrying capacity was investigated in Lanzhou. Experimental results can be summarized that the variation tendency of urban comprehensive carrying capacity was the annual increase of levels and the promotion countermeasure was put forward with congenital restricting factors and acquired restricting factors.

2. Urban Comprehensive Carrying Capacity Index System

2.1. Construction of index system

The social and economic development of all cities are built on the basis of the synergistic effect of many factors, therefore comprehensive index should be adopted to measure the urban comprehensive carrying capacity. Thus on the basis of previous research, and according to the actual situation of the research region, this article will measure the urban comprehensive carrying capacity of Lanzhou city from five aspects including resources carrying capacity, environmental carrying capacity, economic carrying capacity, society carrying capacity, traffic carrying capacity. All of these elements include two indicators, namely the supply of the city for the supporting ability of human activity and the need of the impact of human activities on urban system. The element carrying - capacity reflects the concept of "threshold", and it is characterized by the "short board" in the process of urban development. While the comprehensive carrying- capacity more reflects the concept of "ability", and it is characterized by the "potential" in the process of urban development.

In the process of selecting elements, the principle of being systematic, scientific, operable, flexible and comparable is abide by. The Lanzhou urban comprehensive carrying capacity index system is constructed, including five secondary indexes, namely resources, environment, economy society, and traffic, ten third class indexes such as supply index and demand index, and 20 specific indicators (see figure 1)[1]. The contribution of this work is the society -carrying- capacity which has six indicators, thus breaking the normal situation of only focusing on hard power and ignoring the soft power in such kind of study.

The data of this paper is mainly from the *Lanzhou statistical yearbook from 2003 to 2012, 2003 to 2012, Lanzhou water gazette*, *Lanzhou environment gazettein 2003-2012*, and the traffic data mainly comes from Lanzhou traffic detachment.

2.2. Standardization of evaluation index

The indicators of urban comprehensive - carrying -capacity are different from each other with, large numerical difference. Due to impact of dimensional effect on need of elimination, the original data should be standardized. The 20 specific indicators of the index system can be divided into positive effect indicators and negative effect indicators. The greater the value of the positive effect indicator is, the higher the degree of carrying-capacity is. And the smaller the value of the negative effect indicators is, the higher the degree of carrying-capacity is. Two types of indicators are all processed under range standardization method, but their formulas are different. The specific method is shown as follows:

The normalization equation of the positive effect indicator: $x^*_{ij} = \frac{(x_{ij} - mx_j)}{(Mx_j - mx_j)}$ $i = 1,2,...tn; j = 1,2,...m$;The normalization equation of the positive effect indicator: $x^*_{ij} = \frac{(Mx_j - x_{ij})}{(Mx_j - mx_j)}$ ($i = 1,2,...tn; j = 1,2,...m$) .Among them, Mx_j and mx_j respectively are the maximum value and minimum value of x_j .

2.3. The determination of the weight of index system

The method of determining the weight of the multi-indicators comprehensive evaluation system can mainly be divided into subjective weighting method and objective weighting method[17].The Klee method, Delphi method, AHP method all belong to subjective weighting method. Although this kind of research method is relatively more mature, it has poor objectivity. The advantage of objective weighting method is the calculation based on actual data without relying on persona judgment, thus having strong objectivity.

3. The determination the weight of single indicator

The calculation of the weight of specific indicator in the paper will adopt the mean square error decision-making method of objective weighting method which has relatively clear concept, relatively simple calculation, relatively clear meaning. The basic idea of this method is take the attribute value which is result of the dimensional elimination of scheme A_j under the indicator I_j as the value of the random variable. Firstly calculate the mean square error of all random variables and then unify all these mean square errors. The final results will be the weight coefficients of all single specific indicators. The calculation steps are as follows:

Calculate the mean value of all variables: $\overline{I_j} = \frac{1}{n}\sum_{i=1}^{n} x^*_{ij}$

Calculate the mean square error of $\overline{I_j}$: $\delta(I_j) = \sqrt{\sum_{i=1}^{n} (x^*_{ij} - \overline{I_j})^2}$

The weight coefficient of indicator I_j is : $\omega(I_j) = \delta(I_j) / \sum_{j=}^{m} \delta(I_j)$

3.1. The determination of the weight of the secondary and the third class indicator

After determining the weight of a single specific indicator, multi-objective linear weighting function method is adopted to calculate the attribute value of the first class and secondary indicator . $C_k = \sum x^*_{ij}\omega(I_j)$, $B_j = \sum C_k\omega_k$. C_k for the attribute value of the third class indicator, B_j for the attribute value of the secondary indicator, ω_k for the weight of the third class indicator.

The calculation of the weight of third class indicator adopts the mean square error decision-making method, just like the calculation of the weight coefficient of single indicator. Respectively take the second and third level attribute values B_j and C_k as random variable to calculate the mean square error of the random variable at second and third level, and normalize these mean square errors. The final results will be the weight coefficient of secondary and tertiary indicators, namely ω_i and ω_k.

3.2. The construction of Lanzhou urban comprehensive carrying capacity model

The secondary indicators reflect the development level of urban comprehensive carrying capacity from various factors. The obtained structure of overall level should be comprehensively evaluated. Therefore multi-objective linear weighting function method is adopted to solve the indicator (A) of urban comprehensive carrying capacity, namely $UCCC(A)=\sum_{j=1}^{m}\omega(I_j)x_{ij}^{*}(i=1,2,...tn)$. Finally, the greater the value of UCCC(A)is, the higher the level of urban comprehensive carrying capacity is. Under the circumstance that the weight vector is known, the results over the years can be easily compared according to the value of urban comprehensive carrying capacity and targeting countermeasures and advice can be put forward to improve urban comprehensive carrying capacity.

4. LANZHOU URBAN COMPREHENSIVE CARRYING CAPACITY RESULTS IN 2002-2011

4.1. The contribution of society carrying capacity to urban comprehensive carrying capacity keeps increasing

The urban comprehensive carrying capacity system in Lanzhou is made up of five index system including resources carrying capacity, environmental carrying capacity, economic carrying capacity, society carrying capacity, traffic carrying capacity. The results of calculating weight are: resources carrying capacity is 0.1725, environmental carrying capacity is 0.2025, economic carrying capacity is 0.180, society carrying capacity is 0.2537, traffic carrying capacity is 0.1913. Thus the order of the contributing factors of comprehensive carrying capacity is: society carrying capacity > environmental carrying capacity > traffic carrying capacity > economic carrying capacity >resources carrying capacity. The contributing score of carrying-capacity in every year to comprehensive carrying capacity is shown in figure 2.

Thus, urban development has entered into a stage that both the hard loading factors including resources, environment and the soft loading factors including public services must be taken into consideration. And the city competition in future will mainly focus on soft power.

4.2. Constraint effects of resources carrying capacity keeps strengthening every year

In figure 3, resources carrying capacity shows a decreasing trend year by year, and its constraint effect on urban development tends to strengthen. The reason is that with the development of the city and the increasing pressure of population increase, the contradiction between the supply of urban water resources and land resources and increasing per capita resources utilization becomes more and more prominent with an imbalance between supply and demand. It also can be seen from the data in table 1 that during 2002-2011, the falling range of Lanzhou resources carrying capacity exceeded more than 110% with an average annual decline of 11%.

4.3. The role of environmental carrying capacity has significant changes

In Figure 3 and Table 1, the impact of environmental carrying capacity in 2002 is the lowest, only reaching 0.0481. After treatment, it reached the peak in 2007, and then presented a downward trend year by year, which was closely related with the heavy industry surrounding the Lanzhou city, the rapidly increasing vehicle possessing in the past five years and the frequent hazy weather. The urban environment pressure is great.

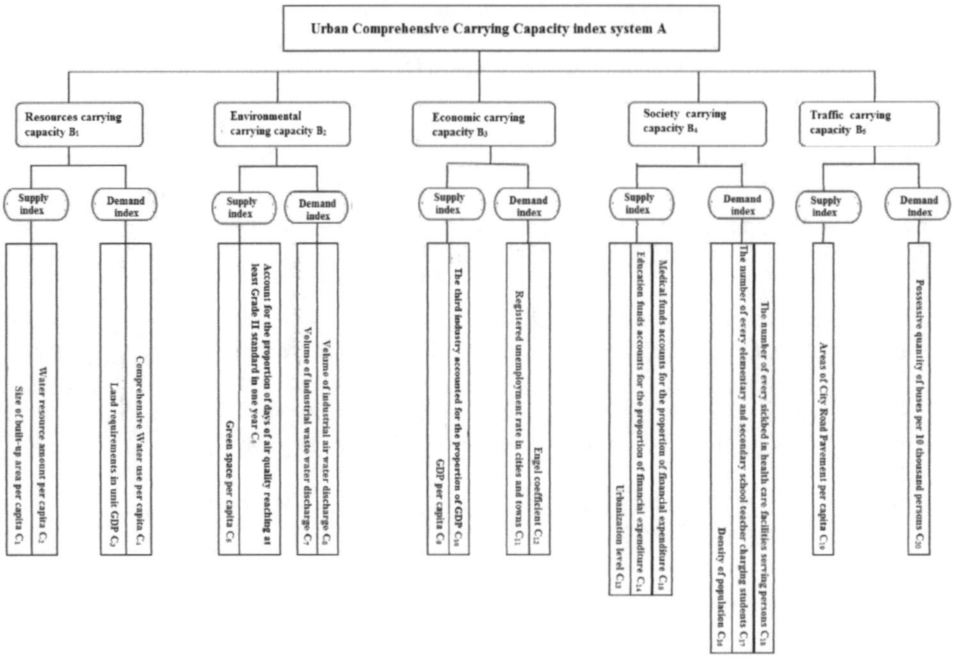

Fig 1 Indicators of urban comprehensive carrying capacity evaluation system

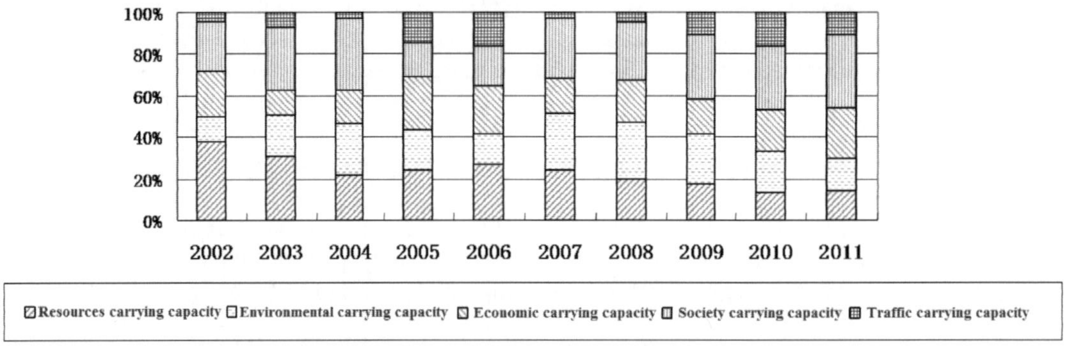

Fig 2 The rate of contribution of factors carrying capacity to urban comprehensive carrying capacity in 2002-2011

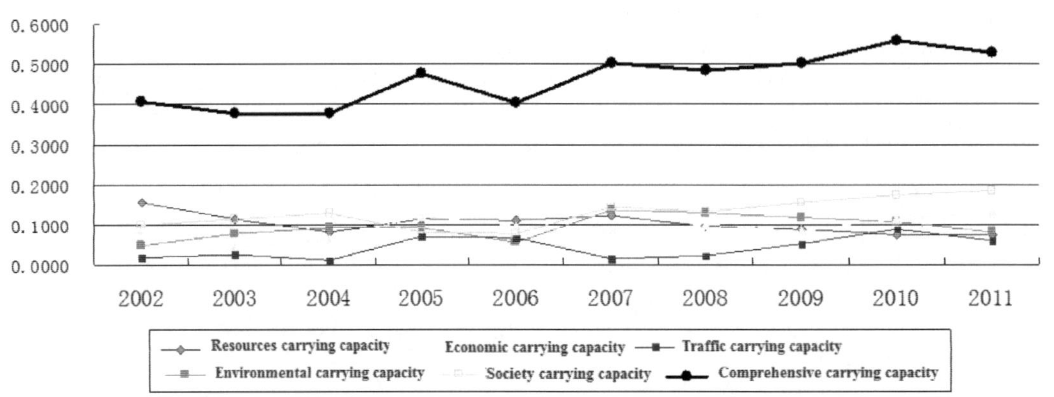

Fig 3 The variation tendency of factors carrying capacity and urban comprehensive carrying capacity in Lanzhou

4.4. Economic carrying capacity shows a steady rising trend

In Figure 3, economic carrying capacity shows a steady rising trend, thus having an more important role in promoting urban comprehensive carrying capacity. Due to the impact of financial crisis, the decline range was larger in 2009. But, generally speaking, the growing rate of the per capita GDP of Lanzhou city in recent three years is great with a significant growth in people's income, economic life and other aspects. And the attraction of being a central city is also strengthening.

4.5. The traffic carrying capacity does not play an important role

In Figure3, the changes of Lanzhou traffic carrying capacity are not regular, sometimes big and sometimes small. Seen from the data provided by Figure 3, the overall level of traffic carrying capacity is not high during ten years. And its peak in 2010 was only 0.0902. The reason is the land form of Lanzhou city: "the east and the west are wide, the north and the south are narrow". Besides, the hinder of the Yellow River also limits the traffic aisle. The urban road capacity is low. Moreover, the exceeding overlap coefficient of bus routes, single public transit system, limiting scope of bus services all lead to great obstacles to the priority of developing public transport.

Table 1 The scores of factors carrying capacity and urban comprehensive carrying capacity in Lanzhou city

Year	2002	2003	2004	2005	2006	2007	2008	2009	2010	2011
Resources Carrying Capacity	0.1564	0.1164	0.0828	0.1159	0.1094	0.1205	0.0962	0.0901	0.0755	0.0741
Environmental Carrying Capacity	0.0481	0.0764	0.0925	0.0910	0.0574	0.1372	0.1314	0.1172	0.1086	0.0820
Economic Carrying Capacity	0.0874	0.0440	0.0617	0.1218	0.0954	0.0868	0.1010	0.0865	0.1143	0.1306
Society Carrying Capacity	0.0988	0.1159	0.1305	0.0797	0.0762	0.1447	0.1342	0.1573	0.1723	0.1855
Traffic Carrying Capacity	0.0170	0.0256	0.0098	0.0692	0.0659	0.0138	0.0225	0.0515	0.0902	0.0575
Comprehensive Carrying Capacity	0.4077	0.3783	0.3775	0.4777	0.4043	0.5029	0.4853	0.5026	0.5609	0.5297

4.6. The urban comprehensive carrying capacity has periodical increase year by year

In Figure 3, Lanzhou urban comprehensive carrying capacity has good development momentum, showing a periodically ascending trend year by year. As a central city in western China and the capital of Gansu province, it should be said that this ascending trend corresponds with the economic and social development trend of the city over the years. The comprehensive carrying capacity from 0.4077 in 2002 to 0.5297 in 2011, with an increase of 29.92% during ten years and an average annual growth of 2.6%.

5. CONCLUSION

The composition of urban comprehensive -carrying- capacity is not the simple sum of the carrying-capacity of various factors but the organic combination of them. At the same time, it also should be recognized that the short board factors of the urban comprehensive carrying capacity reflect in congenital aspect and acquired aspect. And the starting points of breaking through these two constraint factors are different. Congenital short factors is difficult to make a breakthrough, such as land resources, water resources; Acquired short factors have a relative big room to improve, including public services and infrastructure construction which can make a breakthrough through government planning, institutional innovation.

The local resources of all research area are limited. Improving carrying-capacity through increasing the resources needs to break through the congenital constraint factors, which requires for huge cost, and is temporary. Compared with "open source", the starting point of "throttling" based on the demand is more likely to produce results, and are consistent with the spirit of sustainable development. Strengthen the crisis consciousness of the resources and environment from the government, enterprises and individuals at the same time, and establish the idea of sustainable development. The urban development strategy launched by the government should be within the urban carrying-capacity. Beside the reward system of encouraging saving behavior also should be established to guide enterprises and personal resources demand, thus achieving the purpose of "throttling" from the macroscopic aspect. Companies and individuals should be the implementer of "throttling". Enterprises should follow the idea of circular economy, carry out clean production, reduce pollution, and provide green products for the society. Especially, the real estate enterprises should also provide green energy-saving housing for the society, enabling people to save water resource and classify garbage, in daily life, so as to truly achieve "a green four sections ".

Acquired short factors play an important role in improving urban comprehensive carrying capacity. The government can focus on this aspect. The reason of starting from supply is based on urban economic carrying capacity, social carrying capacity and traffic carrying capacity. The government can remove some weak economic increment, diffuse part of population and increase economic vitality through adjusting industrial policy, layout and strengthening. The practice of adjusting and controlling urban scale through adjusting the industrial structure has already been fully understood by many cities and implemented to urban planning and management [4].The improvement of urban-society carrying capacity fundamentally is related with government policy orientation and system innovation. The government should continue to strengthen education, medical and other inputs related with the livelihood of people and provide fair education for the migrant population as well as all kinds of social security. The ascension of urban-traffic-carrying-capacity is to strengthen infrastructure construction and government investment in public transport, set up the new concept of "using public spending to buy public transport service", strive to achieve the new idea arranging prior investment, prior fiscal support, policy support in place, develop the public transport service, so as to further improve the bus trip proportion of citizens.

References

[1] SHI Yishao,YIN Changying,WANG Hefeng,TAN. Research progresses and prospect for Urban Comprehensive Carrying Capacity. Geographical Research.2013(1):133-145.

[2] LIU Xiao;CHEN Jun;FAN Linlin;CUI. Progress and a review of new methods in water resource capacity research. Journal of Beijing Normal University(Natural Science),,2014,50(3) :312-318.

[3] ZHU Yi-zhong, XIA Jun, WANG Gang-sheng. Assessment of water resources carrying-capacity with multi-criteria scenario analysis method:A case study in Zhangye region. Geographical Research, 2005,24(5) :732-740.

[4] Song Xiaomeng,Kong Fanzhe,Zhang Chesheng.Assessment of water resources carrying capacity in Tianjin city of China.Water Resource Management,2011,25:857-873.

[5] YUE Xiao-yan,SONG Ling-ying. Research Review and Prospect of Land Carrying Capacity. Research of Soil and Water Conservation,2008,15(1) :254-257.

[6] GUO Zhiwei. Comprehensive Evaluation on Land Resources Carrying Capacity for Beijing City. Urban Studies,2008,15(5):24-30.

[7] YANG Liang, LV Yao, ZHENG Huayu. Review on Research of Urban Land Carrying Capacity. Progress in Geography,2010,29(5) :593-600.

[8] XU Lin-Yu, YANG Zhi-Feng, LI Wei. Theory and evaluation of urban ecosystem carrying capacity. Acta Ecologica Sinica,2005,25(4):771-777.

[9] WANG Qun, ZHANG Jin-he, YANG Xing-zhu. Analysis of water ecology capacity in Huangshan Resort. Geographical Research,2009,28(4) :1105-1114.
[10] Oh K,Jeonga Y, Leeb D, et al.An integrated framework for the assessment of urban carrying capacity.Korean plan.Assoc,2002,37(5):7-26.
[11] Oha K,Jeonga Y, Leeb D, et al.Determining development density using the urban carrying capacity assessment system.Landscape and Urban planning,2005,73:1-15.
[12] Cang Hui.Carrying capacity,population equilibrium,and enviornment's maximal load.Ecological Modelling 2006,192:317-320.
[13] Lv Bin, SUN Li TAN Wenken, Urban Carrying Capacity Evaluation of Zhongyuan City Agglomeration. China Population Resources and Environment,2008,27(3) :16-20.
[14] Huang Zhiji, Ma Yan He Canfei.Research on Chinese urban agglomerations Carrying Capacity. Urban Problems,2012, (9) :2-8.
[15] HUI Yang-he, JIANG Xiao-hui,HUANG Qiang,XUE Xiao-jie. On system dynamic simulation model of water resources bearing capacity in duality mode. Geographical Research, 2001, 20(2) :191-198.
[16] ZHANG Ke-Yun，FU Shuai-Xiong，ZHANG Wen-Bin. Ecological Carrying Capacity of 31 Provinces Based on Improved Ecological Footprint Model, SCIENTIA GEOGRAPHICA SINICA,2011:31(9) 1083-1089
[17] Wang Mingtao. Multi-index comprehensive evaluation of weights determined deviation, mean square deviation decision method. , CHINA SOFT SCIENCE,.1998(3):100-102

2016 International Congress on Theoretical and Applied Mathematics, Physics and Chemistry IOP Publishing
Journal of Physics: Conference Series **725** (2016) 012005 doi:10.1088/1742-6596/725/1/012005

Wood Texture Features Extraction by Using GLCM Combined With Various Edge Detection Methods

A Fahrurozi[1], S Madenda[1], Ernastuti[1], D Kerami[2]

[1]Computer Science Department, Gunadarma University, Indonesia
[2]Mathematics Department, University of Indonesia, Indonesia

Email: achmad.fahrurozi12@gmail.com

Abstract. An image forming specific texture can be distinguished manually through the eye. However, sometimes it is difficult to do if the texture owned quite similar. Wood is a natural material that forms a unique texture. Experts can distinguish the quality of wood based texture observed in certain parts of the wood. In this study, it has been extracted texture features of the wood image that can be used to identify the characteristics of wood digitally by computer. Feature extraction carried out using Gray Level Co-occurrence Matrices (GLCM) built on an image from several edge detection methods applied to wood image. Edge detection methods used include Roberts, Sobel, Prewitt, Canny and Laplacian of Gaussian. The image of wood taken in LE2i laboratory, Université de Bourgogne from the wood sample in France that grouped by their quality by experts and divided into four types of quality. Obtained a statistic that illustrates the distribution of texture features values of each wood type which compared according to the edge operator that is used and selection of specified GLCM parameters.

1. Background

In an industry that uses wood as raw material, determining the quality of the wood is very important and is known as quality control, because it is useful in the price determination and guarantee the quality of production. In the past, wood quality determination carried by experts through visual observation on a single or several sectional of wood was observed. Along with the development of technology, wood can be observed by using a digital image of a cross section of that wood that forms a particular unique texture. Wood quality is likely to be based on wood fiber density and arrangement of fibers that visualizes the annual rings on that wood. Therefore, the edge detection method used with the hope to strengthen the characteristics contained in image of wood texture. It is because these characteristics can be determined by the fibers contained in wood cross section was observed.

Recognizing of the characteristics of an image of wood texture computationally very useful because it can increase time and cost efficiency in quality control. In order that the computer can recognize the wood characteristics that represented by texture image of wood cross section, it is necessary to extract the characteristics from image of wood texture.These characteristics are usually represented by several items that commonly called features. These features will determine the characteristics of a texture. One method to obtain a texture feature that widely used are Gray Level Co-occurrence Matrices (GLCM) [1]. However in recent years, the GLCM is often combined with other methods and is rarely used individually [1], such as by combining it with Gabor filter [2], Local Binary Pattern (LBP) [3], etc.

Content from this work may be used under the terms of the Creative Commons Attribution 3.0 licence. Any further distribution of this work must maintain attribution to the author(s) and the title of the work, journal citation and DOI.
Published under licence by IOP Publishing Ltd

2. Edge Detection Methods

Edge detection method is used to gain an edge in an image, in which the edge is meant is the boundary between the two areas that have a large enough difference in intensity. Edge detection method used in this study include first-order edge detection (Roberts, Sobel and Prewitt) and second-order edge detection (Canny and Laplacian of Gaussian). First order methods used because their simplicity and second-order used because it is claimed as the most effective edge detection methods (especially Canny), and this research aim to compare the results of first-order and second-order methods.

2.1. First-Order Edge Detection

Roberts edge detection method using a filter with the smallest size ie. 2x2, so it is known as edge simplest edge detection method. While Sobel and Prewitt edge detection method using a 3x3 filter. All of the first-order edge operator provides the results of a grayscale image, where the results of edge detection with Sobel and Prewitt operators tend to have the same characteristics [4].

2.2. Second-Order Edge Detection

Second-order edge detection methods using a filter with varying sizes depending on the values of certain parameters (usually denoted sigma σ). Canny Operator was introduced by John Canny in 1986, and is known as the optimum edge detection operator. Canny edge detection results are a binary image that are proven to capture smooth edges or rough edges in the observed image. While the operator Laplacian of Gaussian (abbreviated as LoG) gives the grayscale or binary image, depending on the size selection and parameter (sigma) in the construction.

According to [5] Canny operator found the edge through six major procedures, ie: 1) Smoothing for noise removal. This can be done by using a Gaussian filter with a simple window. Window used much smaller than the size of the image, 2) Get a edge strength. This is done by using a Gaussian operator, 3) Find gradients use formula *theta* = $\tan^{-1}(G_y, G_x)$, and then convert the *theta* to one of four direction, which is 0°, 45°, 90° and 135° (convertion area shown in figure 1), 4) Non-max suppression 5) Double thresholding and 6) Edge tracking by hysteresis.

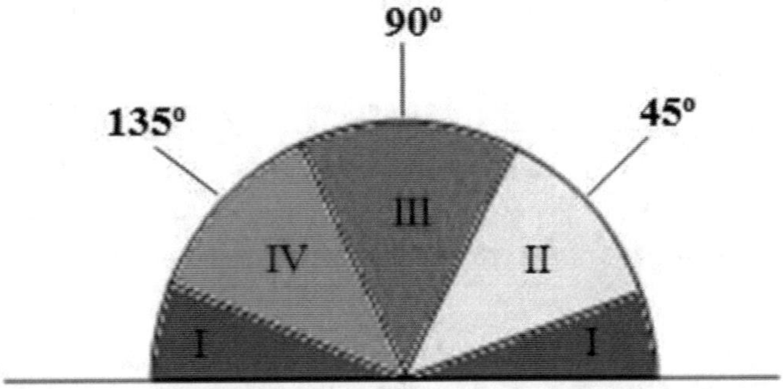

Figure 1. Area convertion of gradients that obtained in step 3). *Tetha* that ranges between 0° and 22,5° well as 157,5° and 180° degrees (area I) converted to 0°, *tetha* that ranges between 22,5° and 67,5° degrees (area II) converted to 45°, *tetha* that ranges between 67,5° and 112,5° degrees (area III) converted to 90°, and *tetha* that ranges between 112,5° and 157,5° degrees (area IV) converted to 135°.

The Laplacian of an image highlights regions of rapid intensity change and is therefore often used for edge detection. The Laplacian is often applied to an image that has first been smoothed with something approximating a Gaussian smoothing filter in order to reduce its sensitivity to noise. The operator normally takes a single gray level image as input [5]. Formula Gaussian functions are used vary, one that is widely used are [6]:

$$LoG(y,x) = -\left[\frac{x^2+y^2-\sigma^2}{\sigma^4}\right] e^{\frac{-(x^2+y^2)}{2\sigma^2}} \quad (1)$$

Journal of Physics: Conference Series **725** (2016) 012005 doi:10.1088/1742-6596/725/1/012005

Meanwhile [8] uses the formula:

$$LoG(y,x) = \frac{1}{\pi\sigma^4}\left[1 - \frac{x^2+y^2}{2\sigma^2}\right] e^{\frac{-(x^2+y^2)}{2\sigma^2}} \quad (2)$$

where σ denote parameter of LoG operator, that can determine size of the filter and how smooth this operator filtering an input image, while *x* and *y* denotes spasial coordinate of any pixels of input image.

3. GLCM

GLCM introduced by Haralick [9], formed by considering the relationship of neighborhood between pixels in a matix, so often used in texture analysis. Adjacency relationship is meant is adjacency between two pixels which are determined by the spatial distance and angles between the two pixels that considered. Thus the spatial distance and angles are parameters of GLCM. Besides the spatial distance and angles, another parameter that determines the construction of GLCM (also called co-occurrence matrice) is a gray limits, which can be interpreted as a parameter that determines the size of the resulting matrix and also determine the rules for conversion of the intensity values for each pixel in original image. The algorithm of co-occurrence matrice construction of a grayscale image can be summarized in figure 2 below.

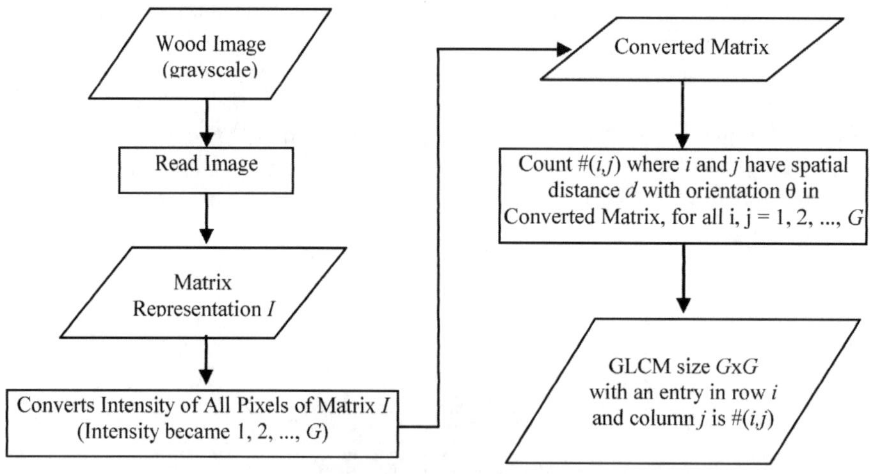

Figure 2. Construction of GLCM from some grayscale image

3.1. Gray Limits Parameter

Parameter Gray limits (denoted *G*) specify the size of the co-occurrence matrix that generated, as illustrated in Figure 3. Selection of parameters *G* determine the rules "grouping" intensity values of the pixels of digital image. Thus the value selection of *G* is based on the characteristics of the given image. Suppose that a grayscale image has an intensity of between 0-255, which means there are 2^8 = 256 gray levels, then the value of *G* that is generally used has the form 2^k, where k = 3, 4, ..., 8 (k =1 for binary image). In general, such an image has $S = 2^m$ gray levels with a matrix representation *I* size *M*x*N* and selected parameter *G* are 2^k, $3 \leq k \leq m$, then the rules of "grouping" the intensity values in matrix *I* are as follows:

1. Determine the divisor factor $p = 2^m/2^k$
2. The Conversion of intensity value for each pixel in the matrix image representation provided that all pixels with intensity values between $h*p$ to $(h+1)*p$-1 converted to (h+1), with h = 0,1,2, …, (G-2), (G.-1).
3. Obtained matrix *I* has an intensity of between 1 to *G*.

Journal of Physics: Conference Series **725** (2016) 012005 doi:10.1088/1742-6596/725/1/012005

The Selection of parameter values for *G* relies on the case at hand and the data they hold. If the data contains imagery that has a texture that is formed due to small differences in intensity between neighboring pixels constituent, then the selection of small value of *G* is feared would "eliminates" the texture of image. So it is better to choose the value of *G* is large enough. However, if the image of textured owned had a very clear edge between the pixels forming the texture, it is possible the selection of *G* value is small, it aims to reduce the size of the resulting matrix GLCM, thus speeding up the process of computin*G*.

3.2. Angle and Direction Parameter

The values of parameter angle (θ) that proposed by Haralick and often used in studies on the analysis and classification of texture are 0°, 45°, 90° and 135°. The determination of adjacency relationship between pixels of the co-occurrence matrix is generated is illustrated in figure 3. For simplicity of writing, the co-occurrence matrices obtained from spatial distance parameter *d* and angles θ denoted $M_{d,\theta}(i,j)$.

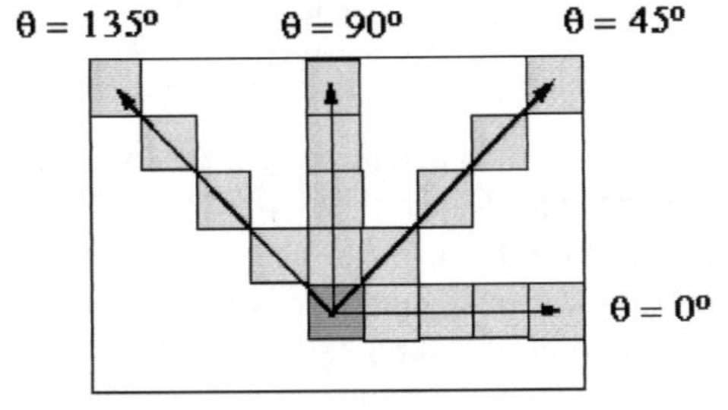

Figure 3. The direction angles that determines the adjacency to obtain ordered pairs of gray level

Another angle that corresponding to the four angles in Figure 4 are 180°, 225°,270° and 315°. The meaning of the corresponding here can be explained as follows: suppose the *d* parameter specified fixed, then the matrix co-occurrence $M_{d,0^\circ}(i,j)$ dan $M_{d,180^\circ}(i,j)$ are matrix that has the entries exactly the same on the main diagonal, and each reflecting the other entries with the main diagonal as axis of symmetry. Mathematically it can be written:

$$m_{d,0^\circ}(i,j) = m_{d,180^\circ}(i,j) \qquad \text{for } i = j$$

$$m_{d,0^\circ}(i,j) = m_{d,180^\circ}(j,i) \qquad \text{for } i \neq j \tag{3}$$

Likewise for another angle that corresponding each other, and because the co-occurrence matrix are a square matrix so we have:

$$M_{d,\theta}(i,j) = (M_{d,(\theta+180^\circ)}(i,j))^T \tag{4}$$

with $(M_{d,(\theta+180^\circ)}(i,j))^T$ denoted the transpose of matrix $M_{d,(\theta+180^\circ)}(i,j)$.

The value of *d* determines the spatial distance to decisive whether the two pixels are neighboring or not. Spatial distance (d) that is used depending on the case at hand, widely 1, 2, 3, 4, 5, and the maximum limits for value of *d* are $L = \min(R\text{-}1, K\text{-}1)$, where *R* is the number of rows and *K* is number of columns of the matrix representation of image I. The smaller the value d, then the adjacency relationships are considered more detail, and vice versa.

Based on the above discussion, if it is determined $G = 8$, the value of $d = 1$, and the direction is determined by the angle $\theta = 0^\circ$, then the entries in the co-occurrence matrix produced was obtained in a manner as illustrated in figure 4 below.

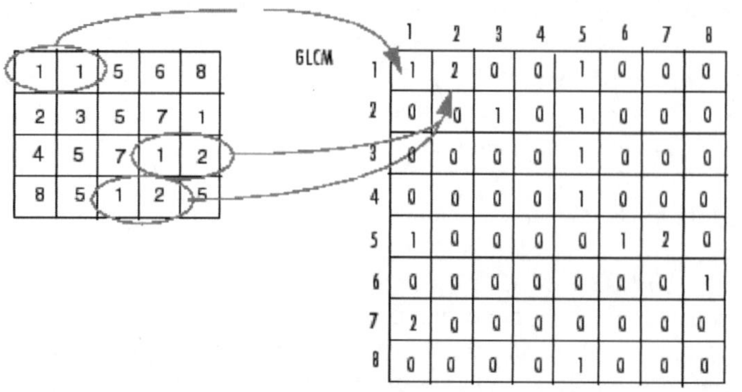

Figure 4. Illustration of co-occurrence matrices construction. Left: Matrix representation of grayscale image I size 5x5 with gray level 1 to 8. Right: Co-occurrence matrice $M_{1,0°}(i,j)$ size 8x8 where the entries $m_{1,0°}(i,j)$ indicates the number of occurrences of gray values i and j being a neighbor with distance $d = 1$ in the direction of 0 ° in the matrix I.

3.3. GLCM Features

Haralick proposed 14 features in "Statistical and structural approaches to texture" in 1979 to characterize texture [9]. GLCM produces features which describe well the relationship of adjacency among pixels in an texture image. These features extracted from a co-occurrence matrices by some formulas, depend on properties that wants to observed. Selection of features to be extracted depends on the case and texture data encountered. Features GLCM frequently used in research and analysis related to the texture of which are Contrast, Correlation, Energy, Entropy, Inverse Difference Moment (IDM) and Homogeneity. In this study, we use four of Haralick texture features, i.e. Contrast, Correlation, Energy and Homogeneity.

3.3.1. Contrast. Contrast is a measure of the intensity contrast between a pixel and its neighbor over the whole image [10]. Contrast is 0 for a constant image [11]. The formula is:

$$\sum_{i=1}^{G}\sum_{j=1}^{G}(i-j)^2 * P_{ij} \tag{5}$$

where P_{ij} denote the probability of an entry $m_{d,\theta}(i,j)$ in GLCM $M_{d,\theta}(i,j)$, and computed with formula:

$$P_{ij} = m_{d,\theta}(i,j) / \sum_{i=1}^{G}\sum_{j=1}^{G} m_{d,\theta}(i,j) \tag{6}$$

3.3.2. Correlation. Correlation measures the linear dependency of grey levels of neighboring pixels. This is often used to measure deformation, displacement, strain and optical flow, but it is widely applied in many areas of science and engineering [12]. Calculated by the formula:

$$\frac{\sum_{i=1}^{G}\sum_{j=1}^{G}(i-1)(j-1)*P_{ij} - \mu_x\mu_y}{\sigma_x\sigma_y} \tag{7}$$

where:
$$\mu_x = \sum_{i=1}^{G}\{(i-1) * \sum_{j=1}^{G} P_{ij}\}$$
$$\mu_y = \sum_{j=1}^{G}\{(j-1) * \sum_{i=1}^{G} P_{ij}\}$$
$$\sigma_x = \sum_{i=1}^{G}\{(i-1-\mu_x) * \sum_{j=1}^{G} P_{ij}\}$$
$$\sigma_y = \sum_{j=1}^{G}\{(j-1-\mu_y) * \sum_{i=1}^{G} P_{ij}\}$$

3.3.3. Energy. Energy is also known as Angular Second Moment or Uniformity, is the sum of squares of entries in the GLCM. Energy measures the image homogeneity. Energy is high when image has very good homogeneity or when pixels are very similar [12]. Energy calculated with the formula:

$$\sum_{i=1}^{G}\sum_{j=1}^{G}\left(P_{ij}\right)^2 \tag{8}$$

where P_{ij} indicated the intensity value of pixel in spatial coordinate (i,j) whole GLCM and G is the value of gray limits parameter of GLCM.

3.3.4. Homogeneity. Homogeneity is meant here is the local homogeneity in the observed image. It is high when local gray level is uniform and inverse GLCM is high [12]. Homogeneity, also known as Inverse Different Moment (IDM), that also large if big values are on the main diagonal. The formula is:

$$\sum_{i=1}^{G}\sum_{j=1}^{G}\frac{P_{ij}}{1+|i-j-2|} \qquad (9)$$

4. Results and Discussion

There are 80 images from 80 samples of wood, which are grouped into four types of quality (determined by expert) and each group consists of 20 samples, provided by LE2i laboratory, Université de Bourgogne, France. The fourth quality of the wood are: (a) very good (type Tres Fin/TF), (b) good (type Fin/F), (c) medium (type Medium/M) and (d) poor (type Gros Grain/GG). The next stage are to apply several methods of edge detection in each image of the sample of wood with the aim of strengthening the appearance of wood fiber that can be used as a measuring tool in determining the quality of the wood. Table 1 below gives an example image of a variety of edge detection that is applied to an image of type F.

Table 1. Image of several edge detection method applied to an image of wood sample from type F.

Original Image	Operator	Edge Detection Image
	Roberts	
	Sobel	
	Prewitt	
	Canny	
	LoG	

Based on the table above, it can be seen that the results of edge detection from several edge detection methods give the expected result, ie strengthen the characteristics of wood image were observed, particularly the fibers that visualizes the annual rings. The results of edge detection using Sobel and Prewitt operator have the same characteristics, while the result of Roberts operator gives the impression that is darker than the other edge detection results. The image produced by Canny operator provides the number of edges that relatively much more than the edges in image produced by using another operators, as mentioned in [5] that this is due Canny can capture the strong and weak edge.

Table 2 below provides the extracted texture features of all types of wood were observed in this study that obtained by using Roberts operator as edge detection method, sets specific GLCM parameters to build co-occurrence matrices the use equation (5), (7), (8) and (9) to compute the texture features.

Table 2. Texture features of Roberts operator image with GLCM parameters $d = 1$, $\theta = 45^{o}$, and $G = 8$

Wood Type	Sample	Features				Wood Type	Sample	Features			
		Contrast	Corr	Energy	Hom			Contrast	Corr	Energy	Hom
TF	01	0.4513	0.2048	0.2408	0.7818	M	01	0.409	0.1944	0.2703	0.7978
	02	0.4263	0.3308	0.2353	0.7964		02	0.3766	0.2813	0.2886	0.8159
	03	0.4522	0.2072	0.2403	0.7816		03	0.4084	0.197	0.2702	0.7981
	04	0.4203	0.3571	0.2356	0.8008		04	0.3402	0.2923	0.3163	0.8313
	05	0.4449	0.2137	0.2421	0.7844		05	0.409	0.1944	0.2703	0.7978
F	01	0.3821	0.1908	0.3034	0.8103	GG	01	0.4214	0.2424	0.2538	0.7966
	02	0.3198	0.2714	0.3683	0.8416		02	0.2805	0.2591	0.4284	0.8604
	03	0.3261	0.2435	0.368	0.8386		03	0.4258	0.2428	0.2532	0.7956
	04	0.3825	0.1856	0.3046	0.8101		04	0.265	0.2532	0.4556	0.8679
	05	0.3877	0.19	0.2994	0.8077		05	0.3861	0.2064	0.2917	0.8083

Suppose it is assumed that the data texture features is the Normal distribution, then the 95% confidence intervals for each values of features obtained can be calculated by the formula [13]:

$$\left[\bar{x} - \left(1.96 * s/\sqrt{n-1}\right),\ \bar{x} + \left(1.96 * s/\sqrt{n-1}\right)\right] \tag{10}$$

where $\bar{x}$, s and n respectively denotes samples means, samples standard deviation, and number of samples. By paying attention to the confidence intervals were obtained, it can be analyzed related to the distribution of texture features values that can be used to distinguish wood type. If the confidence interval for mean of a feature between each type of wood has a fairly large intersection, it can be said that the texture features aren't good enough to used to characterize, or even further to distinguish between one wood texture image and another. Table 3 presents confidence interval for means of the Contrast feature of the whole image of Roberts operator on four types of wood in this paper, where each type has 10 wood texture image.

Table 3. Contrast of Image by Roberts Operator with GLCM parameters: $d = 2$, $\theta = 135^{o}$, and $G = 8$

Sample	Wood Type			
	TF	F	M	GG
01	0.5024	0.4289	0.4513	0.4857
02	0.5452	0.3747	0.4418	0.3311
03	0.5033	0.3739	0.4524	0.4859
04	0.5561	0.4229	0.4207	0.3085
05	0.5003	0.4258	0.4587	0.4311
06	0.5642	0.381	0.4454	0.3436
07	0.5027	0.3758	0.4366	0.4135
08	0.5619	0.4256	0.4376	0.3775
09	0.5337	0.3788	0.4547	0.4881
10	0.2655	0.351	0.4403	0.3383
Mean	0.50353	0.39384	0.44395	0.40033
Std Deviation	0.087667	0.028712	0.011123	0.070116
Int Confidence	0.446255; 0.560805	0.375081; 0.412599	0.436683; 0.451217	0.354521; 0.446139

Due GLCM has some parameters that determine its construction, we then can analyze the effect of changing of GLCM parameters to changes of the distribution of pre-determined texture features. Table 4 presents the effect of changing the spatial distance parameter to confidence interval for the

Energy features in the whole sample of wood (each type of wood has 20 pieces of wood texture images).

Table 4. Confidence interval for means of feature Energy of each type of wood depend on various spatial distance (Fixed parameters $\theta = 0^o$ and $G = 8$, except for operator Canny $G = 2$)

Edge Detection Operator	Wood Type	Spatial Distance 1	2	3	4	5
Roberts	TF	0.2247 ; 0.3110	0.2006 ; 0.2818	0.2000 ; 0.2775	0.2002 ; 0.2750	0.2002 ; 0.2732
	F	0.3503 ; 0.3949	0.3153 ; 0.3574	0.3116 ; 0.3514	0.3090 ; 0.3465	0.3065 ; 0.3426
	M	0.2951 ; 0.3113	0.2645 ; 0.2785	0.2621 ; 0.2749	0.2609 ; 0.2727	0.2596 ; 0.2707
	GG	0.3162 ; 0.4588	0.2845 ; 0.4219	0.2812 ; 0.4140	0.2787 ; 0.4088	0.2766 ; 0.4051
Sobel	TF	0.0306 ; 0.0543	0.0242 ; 0.0444	0.0237 ; 0.0428	0.0238 ; 0.0422	0.0237 ; 0.0415
	F	0.0716 ; 0.0827	0.0598 ; 0.0692	0.0585 ; 0.0674	0.0580 ; 0.0663	0.0573 ; 0.0652
	M	0.0555 ; 0.0598	0.0456 ; 0.0490	0.0446 ; 0.0477	0.0443 ; 0.0473	0.0438 ; 0.0467
	GG	0.0607 ; 0.1001	0.0504 ; 0.0841	0.0494 ; 0.0814	0.0489 ; 0.0800	0.0481 ; 0.0788
Prewitt	TF	0.0565 ; 0.0896	0.0455 ; 0.0743	0.0444 ; 0.0719	0.0444 ; 0.0711	0.0441 ; 0.0702
	F	0.1241 ; 0.1398	0.1057 ; 0.1196	0.1036 ; 0.1168	0.1027 ; 0.1150	0.1016 ; 0.1132
	M	0.0976 ; 0.1042	0.0815 ; 0.0871	0.0796 ; 0.0848	0.0792 ; 0.0842	0.0782 ; 0.0832
	GG	0.1064 ; 0.1604	0.0902 ; 0.1373	0.0884 ; 0.1332	0.0876 ; 0.1313	0.0863 ; 0.1297
Canny	TF	0.4906 ; 0.5093	0.4570 ; 0.4755	0.4569 ; 0.4756	0.4657 ; 0.4850	0.4704 ; 0.4879
	F	0.4757 ; 0.4816	0.4394 ; 0.4448	0.4377 ; 0.4429	0.4449 ; 0.4501	0.4452 ; 0.4506
	M	0.4866 ; 0.5038	0.4517 ; 0.4695	0.4510 ; 0.4691	0.4595 ; 0.4784	0.4633 ; 0.4804
	GG	0.4820 ; 0.5000	0.4464 ; 0.4617	0.4452 ; 0.4593	0.4522 ; 0.4673	0.4533 ; 0.4676
LoG	TF	0.1390 ; 0.1577	0.0961 ; 0.1112	0.0949 ; 0.1107	0.0945 ; 0.1103	0.0941 ; 0.1100
	F	0.1150 ; 0.1189	0.0767 ; 0.0811	0.0741 ; 0.0789	0.0740 ; 0.0787	0.0744 ; 0.0793
	M	0.1233 ; 0.1289	0.0839 ; 0.0889	0.0828 ; 0.0876	0.0821 ; 0.0871	0.0820 ; 0.0872
	GG	0.1156 ; 0.1385	0.0787 ; 0.0971	0.0746 ; 0.0948	0.0749 ; 0.0948	0.0755 ; 0.0956

The effects of changes the parameters spatial distance to a observed feature can be further analyzed by calculating the percentage of intersection of confidence interval for the mean between each types of wood, as presented in table 5. The percentage is calculated by the following formula:

$$PI_{ij} = \frac{I_{ij}}{U_{ij}} \times 100\% \tag{11}$$

with:
PI_{ij} = percentage of intersection of class i and class j,
I_{ij} = length of intersection between the confidence interval for the mean of the data set of the features of class i and class j,
U_{ij} = length of union between the confidence interval for the mean of the data set of the features of class i and class j.

Table 5. Intersection percentage of each confidence interval for mean of feature Energy with fixed GLCM parameters $\theta = 0^o$ and $G = 8$ (except for operator Canny $G = 2$)

	Wood Type	Spatial Distance 1			2			3			4			5		
		TF	F	M	TF	F	M	TF	F	M	TF	F	M	TF	F	M
Roberts	TF	100	-	-	100	-	-	100	-	-	100	-	-	100	-	-
	F	0	100	-	0	100	-	0	100	-	0	100	-	0	100	-
	M	18.29	0	100	17.24	0	100	16.4	0	100	15.75	0	100	15.23	0	100
	GG	0	31.29	0	0	30.61	0	0	29.97	0	0	28.81	0	0	28.07	0
Sobel	TF	100	-	-	100	-	-	100	-	-	100	-	-	100	-	-
	F	0	100	-	0	100	-	0	100	-	0	100	-	0	100	-

	M	0	0	100	0	0	100	0	0	100	0	0	100	0	0	100
	GG	0	28.3	0	0	27.94	0	0	27.9	0	0	26.73	0	0	25.61	0
Prewitt	TF	100	-	-	100	-	-	100	-	-	100	-	-	100	-	-
	F	0	100	-	0	100	-	0	100	-	0	100	-	0	100	-
	M	0	0	100	0	0	100	0	0	100	0	0	100	0	0	100
	GG	0	29.11	0	0	29.3	0	0	29.4	0	0	28.13	0	0	26.92	0
Canny	TF	100	-	-	100	-	-	100	-	-	100	-	-	100	-	-
	F	0	100	-	0	100	-	0	100	-	0	100	-	0	100	-
	M	58.19	0	100	52.65	0	100	49.63	0	100	49.59	0	100	40.83	0	100
	GG	34.47	0	61.52	16.12	0	43.13	7.86	0	34.41	4.83	0	29.78	0	0	15.54
LoG	TF	100	-	-	100	-	-	100	-	-	100	-	-	100	-	-
	F	0	100	-	0	100	-	0	100	-	0	100	-	0	100	-
	M	0	0	100	0	0	100	0	0	100	0	0	100	0	0	100
	GG	0	14.17	24.7	3.16	12.1	27.52	0	20.6	23.82	0.86	18.63	24.88	4.46	17.76	26.01

Based on the calculation results in the foregoing table, shows that the percentage of slices are almost always stable for each operator, except Canny operator. This is presumably because Canny operator generates a binary image with a black background and white edges, and caught a lot of edges that not important so that between the type of wood has almost similar look and co-occurrence matrix. The influence of changing of parameter angle to confidence interval for the mean of the features Energy given in table 6.

Table 6. Intersection percentage of each confidence interval for mean of feature Energy with fixed GLCM parameters $d = 1$ and $G = 8$ (except for operator Canny $G = 2$)

Operator	Wood Type	Angle											
		0			45			90			135		
		TF	F	M	TF	F	M	TF	F	M	TF	F	M
Roberts	TF	100	-	-	100	-	-	100	-	-	100	-	-
	F	0	100	-	0	100	-	0	100	-	0	100	-
	M	18.29	0	100	20.88	0	100	18.83	0	100	13.52	0	100
	GG	0	31.29	0	0	31.79	0	0	32.33	0	0	30.15	0
Sobel	TF	100	-	-	100	-	-	100	-	-	100	-	-
	F	0	100	-	0	100	-	0	100	-	0	100	-
	M	0	0	100	0	0	100	0	0	100	0	0	100
	GG	0	28.3	0	0	29.79	0	0	28.9	0	0	25.64	0
Prewitt	TF	100	-	-	100	-	-	100	-	-	100	-	-
	F	0	100	-	0	100	-	0	100	-	0	100	-
	M	0	0	100	0	0	100	0	0	100	0	0	100
	GG	0	29.11	0	0	31.03	0	0	29.86	0	0	26.4	0
Canny	TF	100	-	-	100	-	-	100	-	-	100	-	-
	F	0	100	-	0	100	-	0	100	-	0	100	-
	M	58.19	0	100	41.37	0	100	16.75	0	100	1.24	0	100
	GG	34.47	0	61.52	1.94	0	36.36	0	9.28	6.31	0	13.08	0
LoG	TF	100	-	-	100	-	-	100	-	-	100	-	-
	F	0	100	-	0	100	-	0	100	-	0	100	-
	M	0	0	100	0	0	100	0	0	100	0	0	100
	GG	0	14.17	24.7	1.19	16.45	30.57	0	16.96	21.58	0	22.27	24.43

Journal of Physics: Conference Series **725** (2016) 012005 doi:10.1088/1742-6596/725/1/012005

Can be seen in the table above that intersection percentage is also quite stable, except Canny operator. This is because the Canny operator produces edges with a random pattern as shown in table 1. While the influence of changing of parameter Gray Limits to confidence interval for the mean of Energy features are presented in table 7. Canny operator is not loaded because Canny generates binary image.

Table 7. Intersection percentage of each confidence interval for mean of feature Energy with fixed GLCM parameters $d = 1$ and $\theta = 0^{o}$

	Wood Type	Gray Limits														
		8			16			32			64			128		
		TF	F	M	TF	F	M	TF	F	M	TF	F	M	TF	F	M
Roberts	TF	100	-	-	100	-	-	100	-	-	100	-	-	100	-	-
	F	0	100	-	0	100	-	0	100	-	0	100	-	0	100	-
	M	19.28	0	100	0	0	100	0	0	100	0	0	100	0	0	100
	GG	0	31.29	0	0	27.62	32.82	0	26.54	0	0	25.54	0.53	0	25.3	0.94
Sobel	TF	100	-	-	100	-	-	100	-	-	100	-	-	100	-	-
	F	0	100	-	0	100	-	0	100	-	4.06	100	-	4.43	100	-
	M	0	0	100	5.56	0	100	16.83	0	100	19.78	0	100	0	0	100
	GG	0	28.3	0	0	27.66	0	0	26.66	0.04	19.95	27.08	0.08	36.17	28.61	1.09
Prewitt	TF	100	-	-	100	-	-	100	-	-	100	-	-	100	-	-
	F	0	100	-	0	100	-	0	100	-	0	100	-	0	100	-
	M	0	0	100	0.07	0	100	2.95	0	100	4.12	0	100	5.14	0	100
	GG	0	29.11	0	0	27.35	0	0	26.99	0	0	26.37	0	0	26.63	0
LoG	TF	100	-	-	100	-	-	100	-	-	100	-	-	100	-	-
	F	0	100	-	0	100	-	0	100	-	0	100	-	0	100	-
	M	0	0	100	0	0	100	0	0	100	0	0	100	0	0	100
	GG	0	14.17	24.7	0	16.45	23.88	0	16.28	23.9	0	16.27	23.79	0	16.21	23.65

In the table above shows that the intersection percentage of Roberts operator and especially Sobel operator suffered considerable changes with increasing values of gray limits parameter than other operators.

5. Conclusion

Based on the experimental results, it was found that edge detection method can strengthen the characteristics of the image of wood texture who want to observed, in particular the appearance of the wood fiber. Feature extraction performed by GLCM provide statistical variation value for each edge detection method that applied. Changes GLCM parameters affect the level differences in the characteristics of the four types of wood that are observed. Based on the features Energy experiments, it was found that greatest affect occur when changes are made in the Graylimits parameter, especially on Sobel operator. A combination that gives the best traits to characterize the type of wood, which is indicated by the number of the smallest intersect (relative to the interval) is Sobel operator and angle parameters. While distributions of data features with worst intersection is a combination of Canny operator and angle parameters. So as to feature Energy, the operator that most appropriate to be used to extract classification feature is the Sobel operator, while the graylimits value used is 8, with the aim of improving accuracy and reducing the size of GLCM used. For other features it is necessary to analyze and further research. The results of this research can then be used to perform classification of four types of wood were observed, by determine the classification method according to the characteristics of the data.

References

[1] Venkataramana, M., *et al.* 2013 *A Review of Recent Texture Classification: Methods* IOSR Journal of Computer Engineering (IOSR-JCE) **14** issue 1 pp. 54-60.

[2] Yosuf, R., Rosli, R., and Khalid, M. 2010 *Using Gabor Filters as Image Multiplier for Tropical Wood Species Recognition System* 12th International Conference on Computer Modelling and Simulation pp. 289-294.

[3] Mahram, A., Shayesteh, M . *G.*, and Jafarpour, S. 2012 *Classification of Wood Surface Defects With Hybrid Usage of Statistical and Textural Features* IEEE Trans. on Systems, Man and Cybernetics, pp. 749-752.

[4] Juneja, M., and Shandu, S. 2009 *Performance Evaluation of Edge Detection Techniques for Images in Spatial Domain* International Journal of Computer Theory amd Engineering **1** pp. 641–621.

[5] Rana, R., and Verma, A. 2014 *Comparison And Enhancement of Digital Image by Using Canny Filter and Sobel Filter* IOSR Journal of Computer Engineering (IOSR-JCE) **16** Issue 1 pp. 06-10.

[6] Crane, R. 1997 *A Simplified Approach to Image Processing Classical and Modern Techniques in C.* Upper Saddle River: Prentice Hall PTR.

[7] Gonzalez R. C., and Woods, R. E. 2008 *Digital Image Processing Third Edition* New York: Pearson Prentice Hall.

[8] Manza, R. R., Gaikward, B. P., and Manza, *G.* R. 2011 *Used of Various Edge Detection Operators for Feature Extraction in Video Scene* UACEE International Journal of Computer Science and Its Applications **2**, Issue 2, pp. 56-60.

[9] V. Bino, Unnikrishnan, A., and Balakrishnan, K., 2012. "Grey Level Co-occurrence Matrices: Generalisation and Some New Features". *International Journal of Computer Science, Engineering and Information Technology (IJCSEIT)*, Vol.2, No.2, pp. 151-157.

[10] Mohan, S., Venkatachalapaty, K., and Priya, S. E. 2014 *Wood Species Identification System* International Journal of Engineering and Computer Science **3** Issue 5 pp. 5996-6001.

[11] Bremananth, R., B., Nithya, and R., Saipriya 2009 *Wood Species Recognition System* International Journal of Computer, Electrical, Automation, Control and Information Engineering **3** No. 4.

[12] P. Mohanaiah, *et al.* 2013 *Image Texture Feature Extraction Using GLCM Approach* International Journal of Scientific and Research Publications **3** Issue 5.

[13] Hogg, R.V. and Craig, A.T. 1995 Introduction *to Mathematical Statistics Fifth Edition* Prentice-Hall International, Inc. New Jersey 07632.

Analysis and Recognition of Curve Type as The Basis of Object Recognition in Image

NURMA NUGRAHA [1], SARIFUDDIN MADENDA [2], DINA INDARTI [3], DEWI AGUSHINTA R [4], ERNASTUTI [5]

E-mail: {[1]nurma, [2]sarif , [3]dina_indarti, [4]dewiar, [5]ernas}@staff.gunadarma.ac.id

Abstract. An object in an image when analyzed further will show the characteristics that distinguish one object with another object in an image. Characteristics that are used in object recognition in an image can be a color, shape, pattern, texture and spatial information that can be used to represent objects in the digital image. The method has recently been developed for image feature extraction on objects that share characteristics curve analysis (simple curve) and use the search feature of chain code object. This study will develop an algorithm analysis and the recognition of the type of curve as the basis for object recognition in images, with proposing addition of complex curve characteristics with maximum four branches that will be used for the process of object recognition in images. Definition of complex curve is the curve that has a point of intersection. By using some of the image of the edge detection, the algorithm was able to do the analysis and recognition of complex curve shape well.

1. Introduction

Digital image processing can be grouped into two types of activities, those are image quality repairing and image information processing. The repairing process of image quality is made to clarify the image so that the image can be viewed without any strain on the eyes and the important information contained in the image can be extracted well. While information processing contained in the image is done for an object recognition by extracting the important information contained in an image. [Marvin., Wijaya., Prijono., A, 2007].

An object in an image when analyzed further will show the characteristics that distinguish one image object with another object. Characteristics that are used in object recognition in an image can be a color, shape, pattern, texture and spatial information that can be used to represent objects in the digital image. A good characteristic is having high distinguishing, so grouping of objects based on the characteristic can be carried out quickly and with high precision.

The method has recently been developed for image feature extraction on the objects that is the curve characteristic similarity analysis and chain code using on object characteristic searching. One is the identification of patterns of squares using eight directions chain code [Ernastuti 2006]. This method successfully characterize the object in the form of plane based on a chaincode that illustrated into a histogram. However, this method has not been able to characterize the object that has the point of intersection.

Image processing research that has not been developed is *Image Recognition Based On Similarity Of Curve Characteristics Invariant To Translation* [Indarti D, Madenda S, Kusuma TM, Agushinta D, 2014]. In that research image characteristics are identified based on the characteristic curve obtained from the edge detection of objects in the image. Characteristic curves used in that research is the type of simple closed curves, simple curves open linear and non-linear open simple curves. The results of image recognition method presented in that research is able to recognize the precise same image based on the similarity of the characteristics of the yield curve edge detection of objects in the image even if the image has undergone a process of transformation. However, that research have not been able to identify the type of complex curve So characteristic curves recognized are still few and limited.
Furthermore, research image recognition using characteristic curve [Indarti, 2014] is applied to an object recognition on the image of ultrasound results are Segmentation Method Development and

Content from this work may be used under the terms of the Creative Commons Attribution 3.0 licence. Any further distribution of this work must maintain attribution to the author(s) and the title of the work, journal citation and DOI.
Published under licence by IOP Publishing Ltd

Shape Feature Extraction Algorithm for Recognition of Breast Tumor type On Ultrasound Image [Lestari DP, 2015]. This research uses eight directions chain code to search contours tumor area. Chain code on the research is used to record the position of each pixel and calculate the number of curves forming tumor area and calculate the code occurrences probability from chain code. The number of curves and the value of the code occurrence probability can be expressed as the difference parameter between one shape of the tumor to another . Curve characteristic that is used in that research is the kind of closed and open simple curves.

Based on the research that has been done before, this research will develop an algorithm analysis and the recognition of the type of curve as the basis for object recognition in images. The image used is the image of the edge detection. Object that can be recognized is a simple object, which has point of intersection the most. This research proposes the addition of the curve characteristic which will be used for the object recognition process on an image, that is, the complex curve characteristic that has maximum four branches. Therefore, curve characteristic which will be used as the basis for object recognition in image in this research is a simple closed curve, linear open simple curves , non-linear open simple curve, the complex closed curve, complex open-linear, complex non-linear curve. By adding the curve characteristic which is used as the basis for object recognition in an image, then the characteristic exists at the object will be more, so grouping of objects based on the features can be done more precisely and accurately.

2. Related Work

2.1. Square Pattern Identification System with Chain Code Method

Method of square pattern identification system with chain code method is divided into training phases and recognition phase. In the training phase. all square image, with a variety of positions, included in the training set, further Canny edge detection method is done. By using detection result is done eight direction chain code methods. From the ranks of chain code formed created histogram. The results of the training phase observation is a pattern of squares, it has the characteristics for each i odd, Hist [i] has equal height or almost equal with the tolerance limit ≤ 1 and for every i is even, Hist [i] has equal height or almost equal to the tolerance limit ≤ 1. Circle pattern and the pattern of equilateral polygons have the feature for all I, Hist [i] has equal height or nearly equal to the tolerance limit ≤ 5. For four rectangle pattern, parallelogram, and diamond has characteristic for i = 1, 5, Hist[i]} has equal height, for i = 2, 6, Hist[i] has equal height, for i = 3, 7, Hist[i] has equal height, and for I = 4, 8, Hist[i] has equal height. , While in irregular sides odd and even patterns have characteristics other than those already discussed.

2.2. Image Recognition Method Based on Similarity of Invariant Curve Characteristic Toward Transformation Geometry

This method is image recognition based on similarity of curve characteristic. The curve character used in this method is simple curve, that is the curve does not have intersection. Stages of research starting from the formation of image database and image recognition. At the formation of image database stage consists of detecting the edge of the image, the curve searching phase, and the recognition curve type stage. In image recognition stage, there is a same stage as database formation until curve type recognition stage. To determine similarity between the characteristic curves in query image were compared with characteristic curve all the images stored in the database. Curve searching is performed pixel per pixel by observing 8 (eight) directions of pixels to determine the position of the next pixel . After all finished curve searching the next step is identify the type of curve that is simple closed curve, linear open simple curves , and non-linear open simple curve. The recognition of the type carried out by characteristics that exist in each curve.

3. Method

Analysis and recognition of curve type as a basic object recognition in image is divided into two stages. Those are complex and simple curves recognition. This paper just focuses on complex curve recognition, that is the curve does not have intersection, the study about simple curve has been done completely by Indarti, 2014. Analysis and recognition of complex curve stage is searching and determining of intersection, determining and searching starting point, curve searching, curve type recognition, and determining of the branch of the curve unity. In generality the algorithm of analysis and recognition of complex curve shown in flowchart at Figure 1.

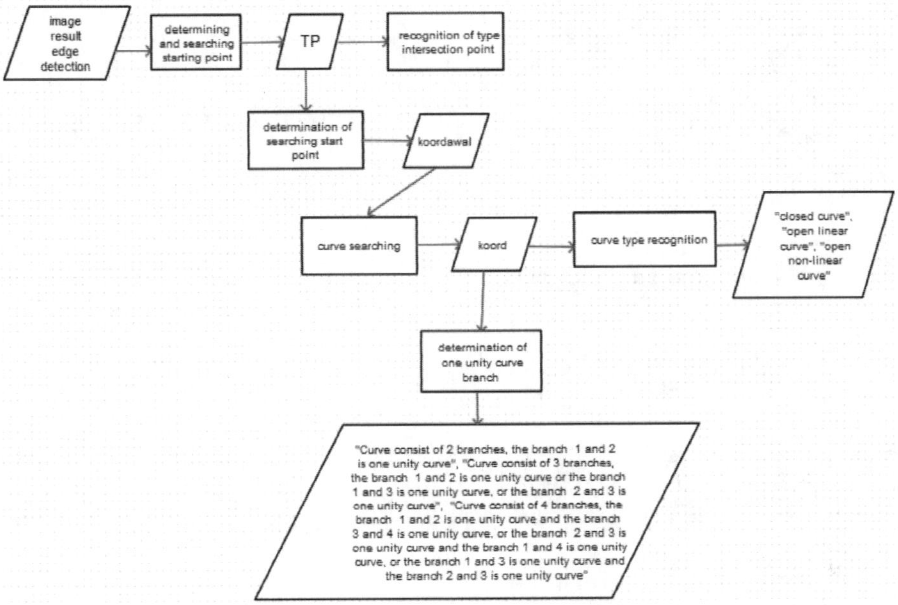

Figure 1. Flowchart algorithm of analysis and recognition of complex curve

3.1. Searching and Determining of Intersection

Two curves said intersect , if two curves lies in a field and have a point of fellowship, communion point is called the point of intersection. The corners formed by the intersection of two curves is the basic idea of searching and determining of the point of intersection of two curves. The angle points of all angels that will be the intersections of two curves. The corners formed by the intersection of two curves could have an angle of 90 °, 45 °, 22.5 °, 11.25 °, and so on. For an idea of searching and determining of intersection point in this research used two linear curves that intersect at an angle of 90 °, 45 °, and 22.5 °. Here is presented a picture of the two curves intersecting at an angle of 90 °, 45 °, and 22.5 °.

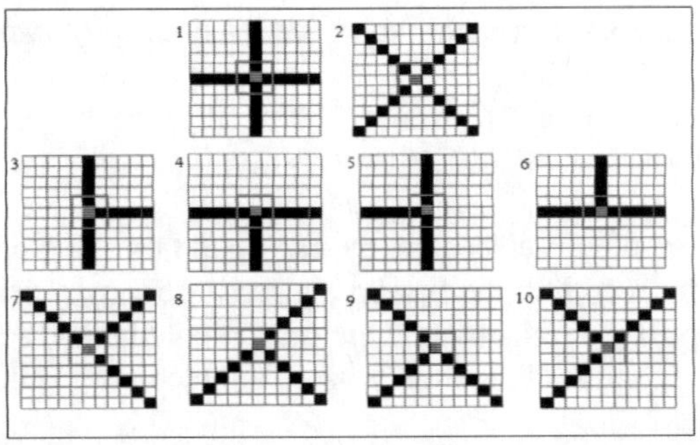

Figure 2. the possibility of two linier curves intersect forming angle 90°

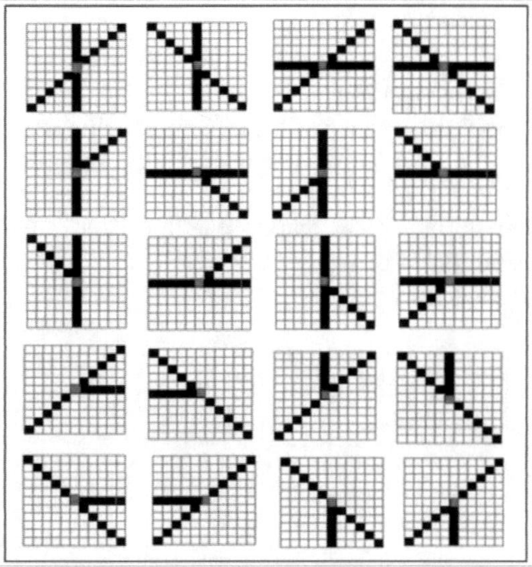

Figure 3. The possibility of two linier curve position intersect forming angle 45°

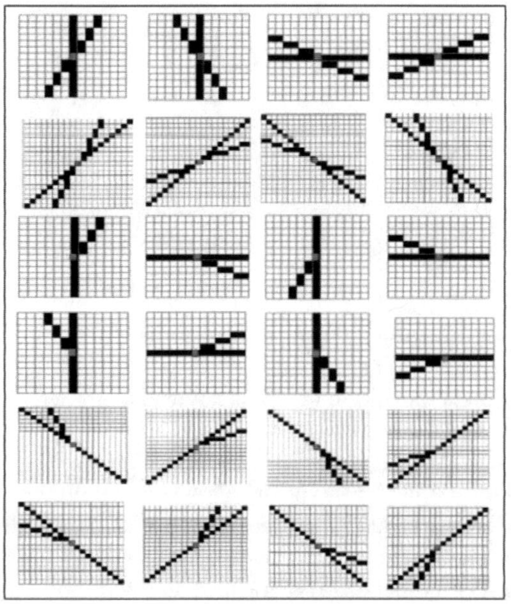

Figure 4. The possibility of two linier curve intersect forming angle 22,5°

Figure 2 shows there are 10 possible positions of the two linear curves that intersect at an angle of 90 °. In Figure 3, there are 18 possible positions of the two linear curves that intersect at an angle of 45 °. While in Figure 4, there are 18 possible positions of the two linear curves that intersect at an angle of 22.5 °. Pixels circled in the picture is the point of intersection or in other words pixels alliance of two intersecting linear curve. Pixel alliance has more than three neighbors. Neighborhoods that should be examined is the environment right and left as well as the top and bottom of the pixel alliance. The examination should be performed on each pixel on the curve.

Search pixel in an edge detection result image in the form of curves is done by tracing the first line of the first column to the first row of the last column. If the search in the first row there is not any pixel that is part of a curve, then continue with the second row of the first column, the second column and so on. For these reasons, the neighborhood that should be examined from every pixel is the environment and the bottom right of each of the pixel. Neighborhoods investigated in the search point on this research are divided into groups. Shown in Figure 5 below.

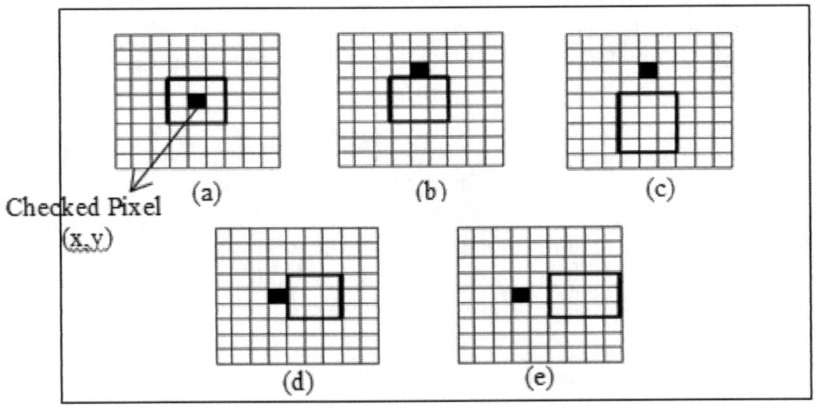

Figure 5 Neighbor Environmental Groups
(a) Group 1, (b) Group 2,
(c) Group 3,
(d) Group 4, (e) Group 5

In Figure 5 (a) group 1 are neighborhoods that are in one column to the right and one column to the left, and one row up and down the rows of pixels are investigated. The number of neighbors (including pixels being investigated) in group 1 was given the name of the NoP. The first group which is most important in determining a candidate point of intersection. In Figure 4 (b) group 2 named NoP 2, Figure 4 (c) 3 groups are named NoP3, Figure 4 (d) group 4 named NoP4, and Figure 4 (e) group 5 named NoP5.

Each pixel part of the curve is checked in environment NoP, NoP2, NoP3, NoP4, and NoP5. If NoP on a pixel amounted to more than 3 then will be further checked. This is because the pixels alliance on a curve has a NoP least more than 3. If a pixel has HoP3, then checking continue to NoP2, NoP3, NoP4, and NoP5. Based on the position of the pixel alliance of two linear curves that intersect at an angle of 90 °, 45 °, and 22.5 ° can be concluded there are 10 conditions that must be met in order for a pixel becomes a cut point. 10 These conditions are shown in Table 1. Based on Table 1 a pixel cutoff point would be candidates if they meet one of the conditions.

Table 1. Type Condition At the candidates of Intersection (x, y)

No	Case	Condition		Intersection
1			NoP2 > 4 NoP3 > 4	$(x+1, y)$
2		NoP > 4	NoP4 > 4 NoP5 > 4	$(x, y+1)$
3	For curves intersecting		NoP2 > 4	$(x+1, y)$
4			NoP2 = 5	$(x+2, y)$
5		NoP = 5	NoP4 = 5	(x, y)
6			NoP2 ≥ 4	$(x+1, y)$
7			-	(x, y)
8			NoP = 4 NoP4 ≥ 4	$(x, y+1)$
9	For curve intercept	NoP = 4	NoP = 4 NoP2 = 3 NoP3 = 3	$(x+1, y+1)$
10		NoP > 3		(x, y)

3.2. The Recognition of Type Intersection

Type of intersection identified in this research is the point of interception to the curve that intersects, and the intersection point for the curves intersect. Intersection point is a point if it is deleted, two linear curves intersect will be divided into four linear curves. Meanwhile intercept point is point if it is deleted, two linear curves intersect will be divided into three linear curves. Examples of curves that intersect and intercept shown in Figure 6.

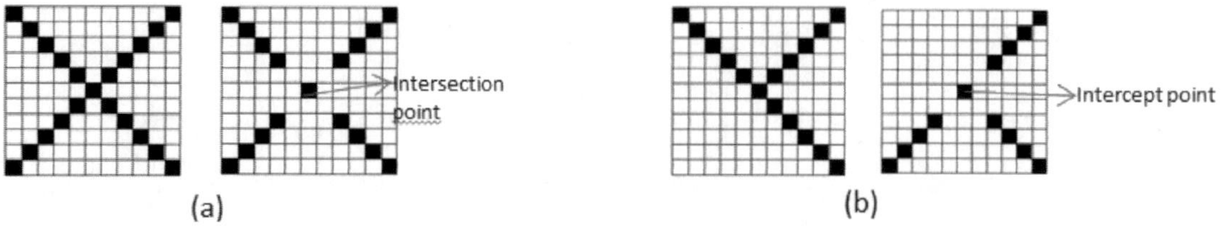

Figure 6 (a) Example curves intersect, (b) Example curves intercept

In Figure 7 (a) clearly shows the intersecting of complex curve, and (b) complex curve intercept , with the intersection point indicated by the arrow. Having obtained the intersection point, the pixels are located around the point of intersection (a pixel above and below, as well as one pixel on the right and left) is removed, it is intended that the pixels of the point of intersection is clearly visible. To determine a curve has a point of interception or intersection point type, defined a neighborhood of the point of intersection has been obtained from the previous step.

Neighborhood 1 is a neighborhood 5x5 or two pixels above and below, as well as two pixels on the right and left from intersection point. Neighborhood 2 is a neighborhood 7x7 or three pixels above and below, as well as three pixels on the right and left from intersection point. Neighborhood 3 is a neighborhood 9x9 or four pixels above and below, as well as four pixels on the right and left from intersection point. Neighborhood 4 is a neighborhood 11x11 or five pixels above and below, as well as five pixels on the right and left from intersection point. All four neighborhoods is shown In Figure 7.

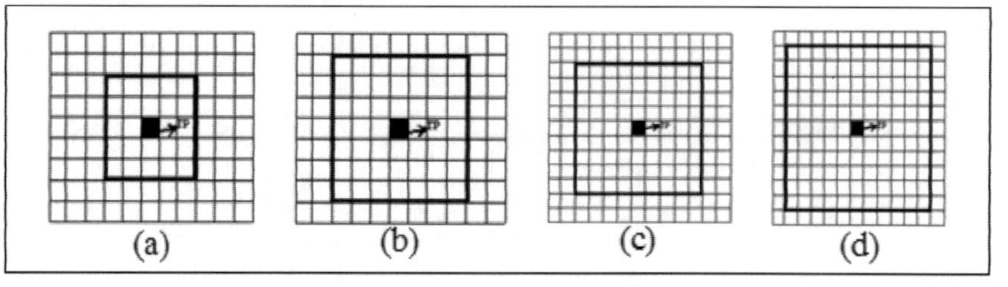

Figure 7 Neighborhood of intersection point
(a) Neighborhood 1, (b) Neighborhood 2, (c) Neighborhood 3, (d) Neighborhood 4

A point called intercept point if the number of black pixel (pixel part of curve) in Neighborhood 1 ≤ 3 pixels, and in Neighborhood 2 ≤ 6 pixels, and in Neighborhood 3 ≤ 9 pixels, in Neighborhood 4 ≤ 12 pixels. If condition does not satisfy, a point called intersection point.

3.3. Determination of Searching Start Point

Searching complex curve is started from intersections, further to ensure that all branches of the curve traced, first determine the starting point of the search. The starting point of the search obtained by identifying the points that are in the neighborhood intersection points. Neighborhood intersection studied is one line upwards and downwards from the point of intersection with the column up and down from the point of intersection. The environment is shown in Figure 8 below.

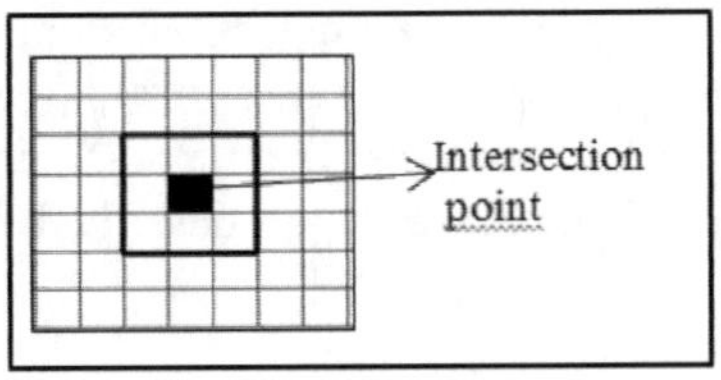

Figure 8. Environment Neighbors Intersection

Chronology of determining the searching starting point starts from counting the number of pixels that are in the neighborhood at the first intersection point on the original curve (not a curve that has eliminated the pixels around the point of intersection). The number of pixels in the neighborhood named NoTP3. If identified NoTP3 $\geq$ 1, look for the location where neighboring pixels of the intersection point by examining 8 neighbors direction. Identifying neighbors of intersection point (x, y) starting from the direction 0, followed by a direction 1, and direction 7, direction 2, direction 6, direction 3, direction 4, and the last direction 5. The layout of the 8 directions neighbor checked are shown in Figure 9 below.

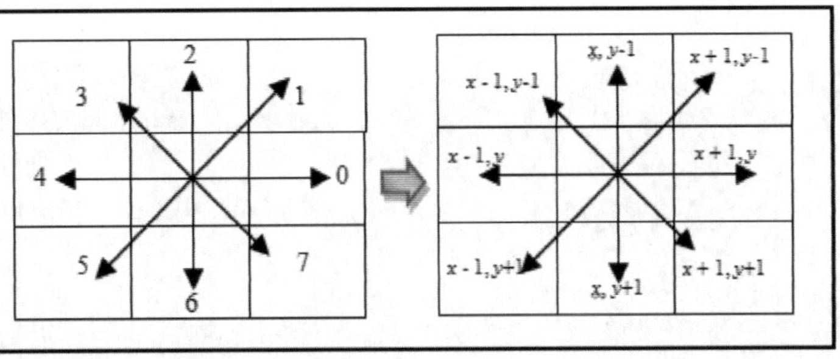

Figure 9 Value Coordinates 8 (eight) Directions of Pixels

On examination NoTP3 if found the existence of the pixels on the environment, then the pixel positions found are stored as the starting point of the branch curve of complex curve. Identification of neighboring pixels from one point of intersection is repeated until the conditions NoTP3 $\geq$ 1 are not satisfied, it means all the neighboring pixels of the first intersection point has been successfully identified. The number of searching start point characterize the number of branches on a curve.

3.4. Curve Searching

Image from edge detection result and vector from the search starting point of curve is used as a search base curve, the search of complex curve starts from the starting point of the first search at the first branch of the first curve. From that point the search is done by checking the neighboring pixel in 8 directions (the same as a check on neighbors intersection point). From the neighboring pixels, checked back the next neighboring pixels, and so on is repeated until all the pixels on the first branch forming of the first curve completely searched. Note that all of the pixels that have been traced is deleted in order to avoid repeated searches in a same curve. The search is performed until all branches traced.

3.5. Curve Type Recognition

The recognition of the curve type performed on each branch at each point of intersection. Each branch of the complex curve recognized its type, is the kind of closed curves, open curve linear, or non-linear open curve. The recognition of the curve type using the data coordinate points of the curve search results. Closed curve is a curve that has a starting point and end point coincident. To determine a branch of the closed curve investigate whether the absolute difference from the starting point and end point of the curve is equal to 0. If yes, this curve is a kind of a closed curve, if not, do further examination.

Further examination are carried out if a curve does not include the type of closed curve, which examined whether the curve is the kind of linear open or non-linear open curve. A curve is said open if the starting point and end point is not coincide. Open curve is divided into two types: linear open curve and non-linear open curve. Discussion on the recognition of the curve types described in [Indarti, 2014].

3.6. Determination of One Unity Curve Branch
After searching and recognizing of curve type at any branch of the curve, the next step is to determine which branches as an unity curve. In determining the curve unity branches used gradient calculation of each branch. Two curve branches that has an equal gradient or nearly equal, then both branches are said to be mutually connecting curve.

4. Result

Given an image is supposed as image the result of edge detection, the image shown in Figure 10. Will be Analyzed and recognition complex curve from that image.

Figure 10. Image of Edge Detection Results Neighbors

In Figure 10 analysis and the recognition of complex curve type. In the image, there are 5 complex curves. The analysis results of the image are:

a. Coordinates (y, x) from the point of intersection and the starting point of the search :
   - Complex Curve 1 : Intersection point at (740.74), and the starting point of the search is (74, 741), (73, 740), (75, 740), (74, 73)
   - Complex Curve 2 : Intersection point at (508.167), and the starting point of the search is (167.509), (166, 508), (168, 508), (167, 507)
   - Complex Curve 3 : Intersection point at (99.197), and the starting point of the search is (197, 100), (196, 99), (198.99), (197, 98)
   - Complex Curve 4 : Intercept point at (633.251), and the starting point of the search is (252, 634), (252.633), (250, 632)
   - Complex Curve 5 : Intersection point at (406.343), and the starting point of the search is (252, 634), (252.633), (250, 632)

b. Results of type of curve recognition

- Complex Curve 1 consists of three branches : Branch 1 and 2 are open non-linear curve, branch 1 is a closed curve
- Complex Curve 2 consists of four branches : The branch 2 and branch 3 is one unity curve and the branch 1 and branch 4 is one unity curve. Branch 2 and 3 are open non-linear curve, Branch 1 and 4 is an open non-linear curve
- Complex Curve 3 consists of four branches : The branch 1 and branch 4 is one unity curve and the the branch 2 and branch 3 is one unity curve. Branch 1 and 4 is an open non-linear curve, branch 2 and 3 are open non-linear curve
- Complex Curve 4 consist of three branches : The branch 2 and branch 3 is one unity curve. Branch 2 and 3 is an open-linear curve, branch 1 is an open-linear curve
- Complex Curve 5 consists of four branches : The branch 2 and 3 is one unity curve and the branch 1 and branch 4 is one unity curve. Branch 2 and 3 are an open-linear curve, branch 1 and 4 are open linear curve.

4. Conclusion

From a number of testing and analysis of the results of each algorithm devised or developed in this research, it can be concluded that the curve type analysis recognition presented in the research consists of several stages and based on the implementation of these methods obtained the following results:

- The algorithms detection and curve searching developed has been able to detect and track the complex curves.
- The curve recognition algorithm created curve can determine the type of the complex.
- The curve recognition algorithm created can define and differentiate between types of open linear and non-linear open curve.

5. Future Work

This research can be developed at recognition of the precise image or the image that has some same characteristics. Furthermore, it can also be developed in the form of the recognition of more complex objects consisting of many intersection point. Applications of this research can also be used on the recognition of the form of letters or numbers.

6. Reference

[1] Canny, J., 1986. "A Computational Approach To Edge Detection". *IEEE Transaction on Pattern Analysis and Machine Intelligence*, 8, pp. 679–698.

[2] Chandwadkar, R., Dhole, S., Gadewar, V., Raut, D., dan Tiwaskar, S.A., 2013. "Comparison of Edge Detection Techniques". *Proceedings of Sixth IRAJ International Conference*, pp. 133-136.

[3] Ernastuti. 2006. "Sistem Identifikasi Pola Bujur Sangkar Dengan Metode Kode Berantai". Majalah Ilmiah Matematika dan Komputer. ISSN 0216-4728. No.3. Universitas Gunadarma

[4] Freeman, H, 1974. "Computer Processing Of Line-Drawing Images". *Computing Surveys,* Vol. 6, No. 1, pp. 57-97.

[5] Lestari, D., P., 2015. "Pengembangan Metode Segmentasi dan Algoritma Ekstraksi Fitur Bentuk Untuk Pengenalan Jenis Tumor Payudara Pada Citra Ultrasonografi". Disertasi, Program Doktor Teknolog iInformasi, Universitas Gunadarma.

[6] Goldstein, L. J, Lay, D. C, dan Schneider, D. I, 2001. *Calculus And Its Applications Ninth Edition.* New York : Prentice Hall.

[7] Gonzalez, R. C. dan Woods, R. E., 2008. *Digital Image Processing Third Edition*. New York : Pearson Prentice Hall.

[8] Gonzales, R., C.,Woods, R., E., Eddins., S., L., 2009. *Digital Image Processing Using Matlab Second Edition*. USA : Gatesmark Publishing.

[9] Indarti, D., 2014. "Pengenalan Citra Berdasarkan Similaritas Karakteristik Kurva Yang Invariant Terhadap Transformasi Geometri". Disertasi, Program Doktor Teknolog iInformasi, Universitas Gunadarma.

[10] Indarti, D., Madenda, S., Kusuma, T.M., Agushinta, D., 2015. "Image Recognation Based On Similarity Of Curve Characteristics Invariant To Translation". *Journal of Theoretical and Applied Information Technology*. Vol 74, No 1.

[11] Indrati, A., 2009. Pengembangan Metode Pengenalan Bentuk dan Batas Tepi Lesi pada Mamografi Digital Sebagai Alat Penunjang Diagnosis Tumor Payudara. Disertasi, Program Doktor Teknolog iInformasi, Universitas Gunadarma.

[12] Jain.,R, Kasturi., R, Schunck., B., G., 1995. *Machine Vision*. McGraw-Hill

[13] Munir, R, 2007. *Pengolahan Citra Digital*. Bandung: Informatika.

[14] Rana, R. dan Verma, A., 2014. "Comparison and Enhancement of Digital Image by Using Canny Filter and Sobel Filter". *IOSR Journal of Computer Engineering,* Vol. 16, Issue 1, pp. 6-10.

[15] Ramadevi Y, Sridevi T, Poornima B, dan Kalyani B. 2010. "Segmentation and Object Recognation Using Edge Detection Techniques". *International Journal of Computer Science & Information Technology (IJCSIT),* Vol 2, No 6.

[16] Shriram K V et. al, 2010. "Automotive Image Processing Technique Using Canny's Edge Detector". *International Journal of Engineering Science and Technology*, Vol. 2 (7), 2632-264

[17] Sutoyo.,T, Mulyanto., E, Suhartono., V, Nurhayati., O., D, Wijanarto, 2009. *Teori Pengolahan Citra Digital*. AndiPublisher.

[18] Marvin., Wijaya., Prijono., A, 2007. *Buku Pengolahan Citra Digital*. Bandung : Informatika.

Dimensionality Reduction of Laplacian Embedding for 3D Mesh Reconstruction

I Mardhiyah[1], S Madenda[2], R A Salim[3], I M Wiryana[4].

[1,2,3,4] Universitas Gunadarma, Depok, Jawa Barat, Indonesia.
Email : {iffatul[1], sarif[2], ravi[3], mwiryana[4]}@staff.gunadarma.ac.id

Abstract. Laplacian eigenbases are the important thing that we have to process from 3D mesh information. The information of geometric 3D mesh are include vertices locations and the connectivity of graph. Due to spectral analysis, geometric 3D mesh for large and sparse graphs with thousands of vertices is not practical to compute all the eigenvalues and eigenvector. Because of that, in this paper we discuss how to build 3D mesh reconstruction by reducing dimensionality on null eigenvalue but retain the corresponding eigenvector of Laplacian Embedding to simplify mesh processing. The result of reducing information should have to retained the connectivity of graph. The advantages of dimensionality reduction is for computational eficiency and problem simplification. Laplacian eigenbases is the point of dimensionality reduction for 3D mesh reconstruction. In this paper, we show how to reconstruct geometric 3D mesh after approximation step of 3D mesh by dimensionality reduction. Dimensionality reduction shown by Laplacian Embedding matrix. Furthermore, the effectiveness of 3D mesh reconstruction method will evaluated by geometric error, differential error, and final error. Numerical approximation error of our result are small and low complexity of computational.

1. Introduction

The 3D shape is a graph which has all the information of vertices. 3D mesh as a surface is viewed by its connected graph. In this case, we discuss about the undirected graph of 3D mesh which has information of the vertices and the faces. Geometric reconstruction is one of application of 3D mesh processing. Due to spectral analysis, geometric 3D mesh for large and sparse graphs with thousands of vertices is not practical to compute all the eigenvalues and eigenvector. Because of that, our new method for 3D mesh reconstruction reduce dimensionality to simplify mesh processing. Reduce dimensionality by reducing spectral will reduce the information, but the result of reducing information should have to retained the connectivity of graph. The advantages of dimensionality reduction is for computational eficiency and problem simplification. Laplacian eigenbases is the point of dimensionality reduction for 3D mesh reconstruction. The connectivity of graph which shows geometric 3D mesh, so one can analysis the geometric and do Laplacian mesh processing. In this paper, we show how to reconstruct geometric 3D mesh after approximation step of 3D mesh by dimensionality reduction. On 3D mesh approximation step, since it will take a lot of time we will not calculate all the information of spectral hereinafter referred dimensionality reduction. Dimensionality reduction shown by Laplacian Embedding matrix. Furthermore, the effectiveness of 3D mesh

Content from this work may be used under the terms of the Creative Commons Attribution 3.0 licence. Any further distribution of this work must maintain attribution to the author(s) and the title of the work, journal citation and DOI.
Published under licence by IOP Publishing Ltd

reconstruction method will evaluated by geometric error, differential error, and final error. Final error is average of geometric error and differential error.
[12] used a small set of eigenvalues and eigenvectors to match 3D shape. Laplacian mesh processing in [12] was done by reducing spectral information but still hold the original 3D mesh geometry. [8] introduce spectral compression of 3D mesh Geometry by spectral method using a linear combination of a number of basis vectors. [2] compressed the 3D triangle mesh geometry on 3D shape geometric representation problem, by projecting to N dimension to build spectral decomposition. They obtained three vectors of dimension N which are called spectra or pseudo frequencies. [13] represents the surface by adding anchors on Laplacian matrix. [11], [13]-[15] introduced the new transformation of Laplacian matrix by differential coordinates called δ −coordinates) and adding anchors on rows of Laplacian matrix. Adding rows on Laplacian matrix made the size of row of matrix growth. Therefore the complexity of their computation was take a lot of time. [3] used TAUCS library [16] as a tools for linear solving on approximation problems. Other methods on 3D mesh reconstruction are Laplace-Beltrami eigenfunctions to understands the geometry of 3D mesh on [10], an ellipsoid shape approximation approach on [9], and Spectral Graph Wavelets (SGW) on [17].

2. Spectral graph

In this study, before 3D mesh reconstruction, we need to graph analysis, dimensionality reduction, and approximate 3D that show on Figure 1. Graph analysis show the geometry of 3D mesh by Laplacian matrix. Laplacian matrix needs vertices location on Euclidean space and connected vertices informations. To simplify a large graph problems which has thousands of vertices, we need to reduce some information but still retain the geometry connectivity, so it is dimensionality reduction purpose on second step. 3D mesh approximation will represent the 3D mesh geometry on other way, so after that we can reconstruct the new 3D mesh.

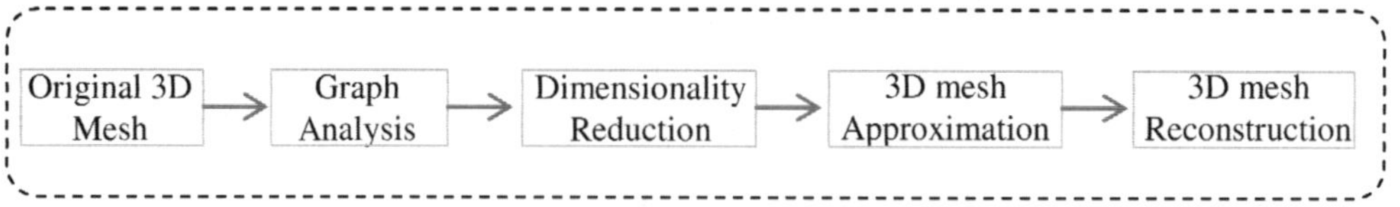

Figure 1. 3D mesh reconstruction steps.

2.1 Graph and the corresponding matrix

Laplacian eigenbases are the important thing that we have to process from 3D mesh information. The information of geometric 3D mesh are include vertices locations and the connectivity of graph. In this paper, $V = \{v_1, v_2, \dots, v_n\}$ on a graph $G(V,E)$ is called by vertices set and the number of vertices is denoted by n. Besides vertices set, on graph $G(V,E)$, set of edge is denoted by $E = \{e_1, e_2, \dots, e_m\}$ and $|E| = m$. The edges shows the connectivity of vertex *i*-th (v_i) and vertex *j*-th (v_j). A graph can be represent by a matrix that we called it as adjacency matrix, it is a symmetric matrix with size $n \times n$, where n is number of vertex in a graph. In general, we can define the matrix adjacency as follow:

$$A = \begin{cases} a_{ij} = 1 & , \text{if there is an edge } e_{ij} \\ a_{ij} = 0 & , \text{if there is no edge} \end{cases} \tag{1}$$

Furthermore, a matrix adjacency can be transformed to weighted matrix by calculating weight for every edge on the graph, weight is calculated using Euclidean distance of v_i (i^{th} vertex) and v_j (j^{th} vertex) which is denoted by c_{ij}. c_{ij} shows the distance of v_i (i^{th} vertex) and v_j (j^{th} vertex), so $c_{ij}^2 = \sum_{i\neq j,i=1}^{n} (v_i - v_j)^2$. Then, weighted matrix can be written as (2):

$$W = \begin{cases} w_{ij} = 1/c_{ij}^2 & \text{, there is an edge between vertex } i^{th} \text{ and vertex } j^{th},\ e_{ij} \\ w_{ij} = 0 & \text{, there is no edge} \end{cases} \tag{2}$$

Let d_v denote the degree of vertex v which represent the number of edge on vertex v. The set of degree of vertex can be showed as a matrix, then it called by degree matrix, denoted by D, $d(v_i) = \sum_{v_i \sim v_j} e_{ij}$, and $D = diag(d(v_1), \dots, d(v_n))$ or $D = \sum_{v_i \sim v_j} w_{ij}$, W is a weighted matrix of graph.

2.2 Graph Laplacian

Graph Laplacian is an important role for 3D mesh reconstruction, so we will see about Laplacian operator. In this paper, 3D mesh geometric reconstruction represented by Laplacian Embedding which the main idea to simplify 3D mesh reconstruction problem. A large graph with many vertices on 3D mesh must be complicated on Laplacian eigenbases calculating problems. Therefore, dimensionality reduction is selected to simplify our problems. 3D mesh reconstruction result must retain the connectivity of graph.

Let $I \in \mathbb{R}^{nxn}$ is identity matrix, $D \in \mathbb{R}^{nxn}$ is degree matrix, $W \in \mathbb{R}^{nxn}$ is weighted matrix, the random-walk Laplacian matrix [4] is given by (3):

$$L = I - D^{-1}W \tag{3}$$

or it can be written as (4),

$$L = \begin{cases} 1 - \frac{w_{ii}}{d_{ii}} & \text{, } i{=}j \text{ and } d_{ii} \neq 0 \\ -\frac{w_{ij}}{d_{ii}} & \text{, there is an edge} \\ 0 & \text{, others.} \end{cases} \tag{4}$$

The weighted matrix and the degree matrix of geometry 3D mesh is shown on (5) with the number of vertex of geometry 3D mesh is n, and the number of faces of geometry 3D mesh is m.

$$\mathrm{W} = \begin{bmatrix} 0 & w_{12} & \cdots & w_{1n} \\ w_{21} & 0 & \cdots & w_{2n} \\ \vdots & \vdots & \ddots & \vdots \\ w_{n1} & w_{n2} & \cdots & 0 \end{bmatrix} \text{ and } \mathrm{D} = \begin{bmatrix} d_{11} & 0 & \cdots & 0 \\ 0 & d_{22} & \cdots & 0 \\ \vdots & \vdots & \ddots & \vdots \\ 0 & 0 & \cdots & d_{nn} \end{bmatrix} \tag{5}$$

2.3 Eigenvalues and Eigenvectors

Vertices on 3D mesh is important to knowing the right positions of the 3D mesh. 3D mesh can represented by vertex and face. The vertices on 3 dimension represented by $3 \times n$ matrix, that rows show XYZ axis of Cartesian coordinates, columns show vertex indices. The face contains indices of vertex that build face in sequence. Laplacian matrix can be obtained by (6)

$$\boldsymbol{Lu} = \lambda \boldsymbol{u} \tag{6}$$

Where, λ is an eigenvalue and $\boldsymbol{u}$ is an eigenvector. For every $f \in \mathbb{R}^n$, we have $f'Lf = \frac{1}{2}\sum_{i,j=1}^{n} w_{ij}^2 (f_i - f_j)^2$, where w_{ij} is scalar. L is symmetric and positive semi-definite, the smallest eigenvalue of L is 0, and the corresponding of eigenvector is the constant one vector. L has non-negative, real valued eigenvalues $0 = \lambda_1 \leq \lambda_2 \leq \cdots \leq \lambda_n$. A is a symmetric matrix which shows the connectivity of all the vertices from the connected graph. A called by adjacency matrix. 3D mesh with n vertices has n real eigenvalues and n real eigenvectors form an orthonormal basis. $\{\lambda_1, \lambda_2, \dots, \lambda_n\}$ are eigenvalues of the adjacency matrix and $\{\boldsymbol{u}_1, \boldsymbol{u}_2, \dots, \boldsymbol{u}_n\}$ are eigenvectors. All the eigenvectors corresponding with eigenvalues. The eigenspace $S_i = \{\boldsymbol{u} \in \mathbb{R}^n | A\boldsymbol{u} = \lambda_i \boldsymbol{u}\}$ where $1 \leq i \leq n$ represent set of the eigenvectors [7]. Then (7), $\boldsymbol{\Lambda}$ shows set of the eigenvalue of L and $\boldsymbol{U}$ shows set of the eigenvectors.

$$\boldsymbol{\Lambda} = \begin{bmatrix} \lambda_1 & 0 & \cdots & 0 \\ 0 & \lambda_2 & \cdots & 0 \\ \vdots & \vdots & \ddots & \vdots \\ 0 & 0 & \cdots & \lambda_n \end{bmatrix} \text{ and } \boldsymbol{U} = \begin{bmatrix} u_1(v_1) & u_2(v_1) & \cdots & u_n(v_1) \\ u_1(v_2) & u_2(v_2) & \cdots & u_n(v_2) \\ \vdots & \vdots & \ddots & \vdots \\ u_1(v_n) & u_2(v_n) & \cdots & u_n(v_n) \end{bmatrix} \quad (7)$$

2.4 Dimensionality Reduction

In this paper, we will reduce the dimension of Laplacian matrix by reducing the dimensionality of spectral graph. Spectral dimensionality reduction step is for reducing the number of spectral on 3D mesh approximation. The reduction of computational of approximation should retains the connectivity graph. The 3D mesh geometric reconstruction results must be show the 3D mesh approximation that represent the original 3D mesh. By embedding the graph into the spectral space, we can get 3D mesh reconstruction. By projecting all vertices in the graph into k dimension of spectral space. Graph that represent 3D mesh as a geometric form is a finite graph. On [1], every finite graph can be embedded in k Euclidean space $(\mathbb{R}^k)$ if $k \geq 3$. We use k spectral information of 3D mesh for 3D mesh reconstruction purposing. k is the independent variable, denotes how many spectral information that used for 3D mesh reconstruction purposing by spectral dimensionality reduction. Because of reducing dimension (spectral), it will makes efficient on computational 3D mesh reconstruction processing.

Assume that we have one connected graph of 3D mesh. 3D mesh have the Laplacian matrix and we can get the eigenvalue matrix and eigenvector matrix. Υ shows the original eigenvalues matrix after spectral dimensionality reduction by k information on (8).

$$\Upsilon = \begin{bmatrix} \lambda_1 & 0 & \cdots & 0 \\ 0 & \lambda_2 & \cdots & 0 \\ \vdots & \vdots & \ddots & \vdots \\ 0 & 0 & \cdots & \lambda_k \end{bmatrix} \quad (8)$$

And U shows the original eigenvectors matrix after spectral dimensionality reduction by k information on (9).

$$\mathrm{U} = \begin{bmatrix} u_1(v_1) & u_2(v_1) & \cdots & u_k(v_1) \\ u_1(v_2) & u_2(v_2) & \cdots & u_k(v_2) \\ \vdots & \vdots & \ddots & \vdots \\ u_1(v_n) & u_2(v_n) & \cdots & u_k(v_n) \end{bmatrix} \quad (9)$$

We obtain $n \times k$ matrix which $\boldsymbol{u}_i^T \boldsymbol{u}_j = \alpha_{ij}$ (orthonormal vectors, $i, j = 1, \dots, k$), hence $U^T U = I_k$. The eigenvalue of Laplacian matrix is called by spectral Laplacian. The spectral Laplacian considered approximate the mesh using linear combination of a number of basis vectors. On [6] $\boldsymbol{u}_2$ is called Fiedler vector.The first eigenvalue is zero that corresponding by first eigenvector $\boldsymbol{u}_1 = [1,1,1, \dots ,1]$ (size n). Eigenvalues of Laplacian matrix corresponding by geometry of 3D mesh are called spectral graph for 3D mesh. The first eigenvalue (λ_1) is 0, the first column of eigenvector reduced and we have $\{\boldsymbol{u}_1, \boldsymbol{u}_2, \dots, \boldsymbol{u}_k\}$ as an eigenvector. Therefore, to build Laplacian Embedding, spectral dimensionality reduction of Laplacian eigenbases by k information are shown on (8).

3. 3D Mesh Reconstruction

In the 3D mesh reconstruction, we approximate all vertices to get the new vertices which will reconstruct 3D mesh. We project the original geometry of 3D mesh into $k - 1$ dimension, and vertices approximation obtained by back projection into 3 dimension on approximation stage. The vertices approximation are reconstructed. Inverse the L-embedding is our work to reconstruct 3D mesh so that we should not have to adding anchors to the original Laplacian matrix like [15]. The set V (vertices) represents the geometrical information of the mesh and the set E (edges) represents the connectivity information. $\boldsymbol{v} = (\boldsymbol{x}, \boldsymbol{y}, \boldsymbol{z}) \subset \mathbb{R}^3$ in Cartesian coordinates shows the geometry of 3D mesh. The 3D mesh embed into $k-1$ dimension, so we obtain three vectors of $k-1$ dimension which called spectra or pseudo-frequencies $(\boldsymbol{x_p}, \boldsymbol{y_p}, \boldsymbol{z_p})$ as [2]. $(\boldsymbol{x_p}, \boldsymbol{y_p}, \boldsymbol{z_p})$ correspond to the 3D mesh original vertices. Since λ_1 is zero, λ_1 is eliminated from the eigenvalues matrix. On this 3D mesh reconstruction, the embedding method is reducing dimensionality by eliminate null eigenvalue (first eigenvalue) but retain the corresponding eigenvector (first eigenvector). Besides that, rearrange the column of eigenvectors matrix with the result that $\{\boldsymbol{u}_k, \boldsymbol{u}_{k-1}, \dots, \boldsymbol{u}_1\}$ and eliminated the first column, so eigenvector $\boldsymbol{u}_k$ is eliminated. (10) shows the new eigenvalues matrix, and (11) shows the new eigenvector matrix.

$$Y = \begin{bmatrix} \lambda_2 & 0 & \cdots & 0 \\ 0 & \lambda_3 & \cdots & 0 \\ \vdots & \vdots & \ddots & \vdots \\ 0 & 0 & \cdots & \lambda_k \end{bmatrix} \tag{10}$$

$$M = \begin{bmatrix} u_{k-1}(v_1) & u_{k-2}(v_1) & \cdots & u_1(v_1) \\ u_{k-1}(v_2) & u_{k-2}(v_2) & \cdots & u_1(v_2) \\ \vdots & \vdots & \ddots & \vdots \\ u_{k-1}(v_n) & u_{k-2}(v_n) & \cdots & u_1(v_n) \end{bmatrix} \tag{11}$$

The number of eigenfunctions must be $3 \leq k \leq n-1$ which can be implied to embed the graph. The Laplacian Embedding of a graph as an operator to embed the original 3D mesh into $k-1$ dimension can be viewed as (12).

$$\ell = Y^{-1/2} M^T \tag{12}$$

Therefore, the embedded coordinates of a vertex v_j are (13):

$$\ell_j = \begin{pmatrix} \frac{u_{k-1}(v_j)}{\sqrt{\lambda_2}} \\ \vdots \\ \frac{u_1(v_j)}{\sqrt{\lambda_k}} \end{pmatrix}. \tag{13}$$

The projected vertices of k-dimension can be obtained by multiplying the Laplacian Embedding matrix with the coordinates (14):

$$\begin{aligned} \boldsymbol{x_p} &= \boldsymbol{\ell x} \\ \boldsymbol{y_p} &= \boldsymbol{\ell y} \\ \boldsymbol{z_p} &= \boldsymbol{\ell z} \end{aligned} \tag{14}$$

The Laplacian embedding is not symmetric matrix, so we calculate its pseudo-inverse for equation (12) to get the approximation of 3D mesh information. The new vertices or an exact reconstruction are $\widehat{\boldsymbol{x}} = \boldsymbol{x}_p^T \boldsymbol{\ell}^+$, $\widehat{\boldsymbol{y}} = \boldsymbol{y}_p^T \boldsymbol{\ell}^+$, $\widehat{\boldsymbol{z}} = \boldsymbol{z}_p^T \boldsymbol{\ell}^+$. Where $\boldsymbol{\ell}^+$ represents the pseudo-inverse of (12). That new vertices are used to reconstruct 3D mesh, such that the numerical approximation error of 3D mesh reconstruction are small and low complexity of algorithm.

4. Experimental Results

In this study, we show three kinds of 3D mesh that has 4315, 5103, and 24955 vertices. We use OFF file format for vertices and faces detail information. In Figure 2 we approximate the 3D mesh of cow. The geometry of 3D mesh cow has 4315 vertices and 8626 faces, the original mesh shows in Figure 2.

Journal of Physics: Conference Series **725** (2016) 012007 doi:10.1088/1742-6596/725/1/012007

Figure 2. Cow original mesh has 4315 vertices.

The 3D mesh geometric reconstruction of cow by reducing spectral $k = 216, 863, 1726,$ and 2158. These k parameters are equal to 5%, 20%, 40%, and 50% as approximation ratio. Therefore, the reconstruction of 3D mesh of cow results show in Figure 3.

(a) (b) (c) (d)

Figure 3. Geometric reconstructions of 3D mesh cow with k=216 (a), 863 (b), 1726 (c), and 2158 (d) the increasing number of eigenfunctions.

Hereafter we will see the other 3D mesh, such as homer and elephant. Homer has 5103 vertices and 10202 faces. Figure 4 shows the original 3D mesh of homer with 5103 vertices.

Figure 4. Homer original mesh has 5103 vertices.

The 3D mesh geometric reconstruction of homer by reducing spectral $k = 1021, 2041,$ and 2552. These k parameters are equal to 20%, 40%, and 50% as approximation ratio. Then, the reconstruction of 3D mesh of homer results show in Figure 5.

(a) (b) (c)

Figure 5. Geometric reconstructions of 3D mesh homer with k=1021 (a), 2041(b), and 2552 (c) the increasing number of eigenfunctions.

Figure 6 shows the original 3D mesh of elephant that has 24955 vertices and 49918 faces.

Figure 6. Elephant original mesh has 24955 vertices.

The 3D mesh geometric reconstruction of elephant by reducing spectral $k = 250, 499, 749$ and 998. These k parameters are equal to 1%, 2%, 3%, and 4% as approximation ratio. Then, the reconstruction of 3D mesh of homer results show in Figure 7.

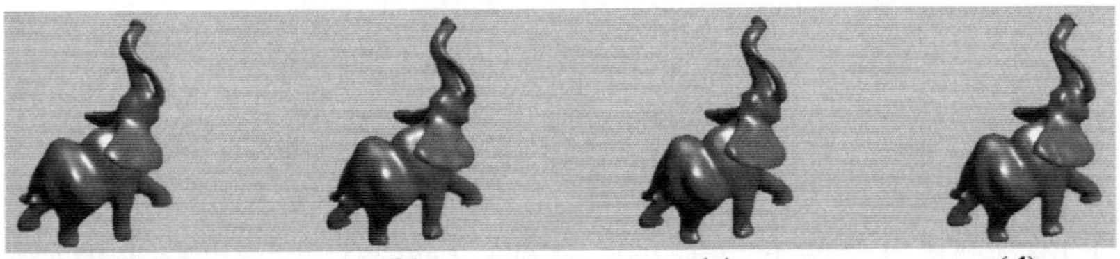

(a) (b) (c) (d)

Figure 7. Geometric reconstructions of 3D mesh elephant with *k*=250 (a), 499 (b), 749 (c), and 998 (d) the increasing number of eigenfunctions.

These reconstruction results are closely enough to described the original 3D mesh as the approximation of all 3D mesh. The connected graph on the 3D mesh are retained on 3D mesh reconstruction.

4.1 Evaluation methods

On 3D mesh geometric reconstruction, certainly there is error on approximation results, since generated 3D mesh is just simply a fiction of the original object. [15] evaluate the approximation result by metro tool [5]. In this paper discussed error calculation that occurred in the approximation of geometric 3D mesh with three kinds of error. Three kinds of error are geometric error, differential error, and final error. On error calculation, there are M_1 which represents the original 3D mesh, while M_2 represents the approximation 3D mesh. (15) shows geometric error formula.

$$g_e = \|M_1 - M_2\|_g = \sum_{i=1}^{n} \frac{1}{n} \left\| v_i^1 - v_i^2 \right\|_2 \tag{15}$$

with

n shows the number of 3D mesh vertices

v_i^1 shows the original vertex of 3D mesh (v) i^{th}, $1 \le i \le n$,

v_i^2 shows the original vertex of 3D mesh $(\hat{v})$ i^{th}, $1 \le i \le n$.

[8] introduce Geometric Laplacian (GL) then calculate differential error between both vertex (original vertex and approximation vertex) that showed on (16).

$$GL(v_i) = v_i - \frac{\sum_{j \in N(i)} l_{ij}^{-1} v_j}{\sum_{j \in N(i)} l_{ij}^{-1}}, \tag{16}$$

l_{ij}^{-1} shows inverse of the edge length between v_i and v_j, such that differential error can be written as (17).

$$d_e = \|M_1 - M_2\|_d = \sum_{i=1}^{n} \frac{1}{n} \left\| GL(v_i^1 - v_i^2) \right\|_2 \tag{17}$$

[17] used evaluation method of their compression results by the average of geometric error and differential error which are called final error (fe) on (18).

$$fe = \|M_1 - M_2\| = \frac{1}{2}\left(\|M_1 - M_2\|_g + \|M_1 - M_2\|_d\right) \tag{18}$$

All our experiments run on computer with core i7, 2.20 GHz, and RAM 4 GB as specifications. To compute the eigenvalue and the eigenvector, we used MATLAB. We compare three kinds of 3D mesh in this study, such as cow, homer, and elephant that are .OFF file format. On .OFF file format, one can know about the number of vertices of the object of 3D mesh. Table 1 shows error approximation for 3D mesh cow reconstructions, which are showed in Figure 2 with 4315 vertices by some ratio, 5%, 20%, 40%, and 50%. The evaluation dimensionality reduction method of 3D mesh cow reconstruction by Laplacian Embedding include geometric error, differential error, and final error.

Table 1. Error approximation for 3D mesh cow with 4315 vertices.

Ratio	Geometric Error	Differential Error	Final Error
5%	1.47e-4	2.24e-4	1.86e-4
20%	0.94e-4	1.47e-4	1.21e-4
40%	1.08e-4	1.31e-4	1.19e-4
50%	1.05e-4	1.20e-4	1.12e-4

Table 2 shows error approximation for 3D mesh homer reconstructions, that are showed in Figure 4 with 5103 vertices by some ratio 20%, 40%, and 50%. The evaluation dimensionality reduction method of 3D mesh homer reconstruction include geometric error, differential error, and final error.

Table 2. Error approximation for 3D mesh for 3D mesh homer with 5103 vertices

Ratio	Geometric Error	Differential Error	Final Error
20%	2.62e-5	2.50e-5	2.56e-5
40%	2.16e-5	2.01e-5	2.08e-5
50%	3.38e-5	2.89e-5	3.13e-5

Table 3 shows error approximation for 3D mesh elephant reconstructions, which are showed in Figure 6 with 24955 vertices by some ratio, 1%, 2%, 3%, and 4%. The evaluation dimensionality reduction method of 3D mesh elephant reconstruction include geometric error.

Table 3. Error approximation for 3D mesh elephant

Ratio	Geometric Error
1%	1.75e-5
2%	1.26e-5
3%	8.74e-6
4%	7.60e-6

5. Conclusions

We have developed the new method to 3D mesh reconstruction which derive basis from mesh connectivity by reducing dimensionality on null eigenvalue but retain the corresponding eigenvector. The main idea is dimensionality reduction of Laplacian Embedding by reducing any spectral but retain the connectivity of graph as 3D mesh. Built the Laplacian embedding by eliminating the first eigenvalue, rearrange the column of eigenvector matrix, and eliminating the first column of the rearrange eigenvector matrix result. By embedding all the original vertices, the approximation vertices can be obtained by multiplying the projected vertices and pseudo-inverse of Laplacian embedding. The result of dimensionality reduction on 3D mesh reconstruction is for simplify reconstruction problems and computational eficiency. Since reducing dimension (k spectral) which $3 \leq k \leq n-1$, we made efficient on computational by Laplacian embedding without adding anchors on 3D mesh geometric reconstruction processing.

Journal of Physics: Conference Series **725** (2016) 012007 doi:10.1088/1742-6596/725/1/012007

References

[1] Brouwer, A.E., Haemers, W.H. 2011. *Spectra of Graph*. New York : Universitext-Springer

[2] Cayre, F., Rondao-Al*face*, P., Schmitt, F., Macq, B., Maître, H. 2003. *Application Of Spectral Decomposition To Compression And Watermarking Of 3D Triangle Mesh Geometry*. Elsevier Science, **18**:309-319.

[3] Chen, D., Cohen-Or, D., Sorkine, O., Toledo, S. 2005. *Algebraic Analysis of High-pass Quantization*, ACM Transaction on Graphics, Vol.24(4), pp. 1259-1282.

[4] Chung, F.R.K. 2000. *Spectral Graph Theory*. CBMS. American Mathematical Society.

[5] Cignoni, P., Rocchini, C., Scopigno, R. 1998. Metro : *Measuring Error on Simplified Surfaces*. Computer Graphics Forum, Vol. 17, no. 2, pp. 167-174.

[6] Fiedler, M. 1973. *Algebraic Connectivity of Graphs*. Czechozlovak Mathematical Journal, Vol. 23, No. 2, pp. 298-305.

[7] Horaud, R. 2012. *A Short Tutorial On Graph Laplacians, Laplacian Embedding, And Spectral Clustering*. INRIA.

[8] Karni, Z., Gotsman, C. 2000. *Spectral Compression of Mesh Geometry*. Proceedings of ACM SIGGRAPH 2000, pp. 279-286.

[9] Khatun, A., Chai, W. Y., Iskandar, DNF. A., Islam, M.R. 2012. *An Ellipsoid Shape Approximation Approach for 3D Shape Representation*. Proceeding of the IIEEJ Image Electronics and Visual Computing Workshop. Kuching, Malaysia.

[10] Lévy, B. 2006. *Laplace-Beltrami Eigenfunctions Towards An Algorithm That "Understands" Geometry*. INRIA.

[11] Lipman, Y., Sorkine, O., Cohen-Or, D., Levin, D., Rössl, C., Seidel, H.P. 2004. *Differential Coordinates for Interactive Mesh Editing*. Proceedings of EUROGRAPHICS/ACM SIGGRAPH Symposium on Geometry Processing (SGP), pp. 42-51.

[12] Sharma, A., Horaud, R., Mateus, D. 2012. Chapter 13 : 3D shape registration using spectral graph embedding and probabilistic matching. *Image Processing and Analysis with Graphs : Theory and Practice*. CRC Press

[13] Sorkine, O. 2005. State-of-the-art : Laplacian Mesh Processing. *Proceeding of EUROGRAPHICS.*

[14] Sorkine, O. 2006. *Differential Representations For Mesh Processing*. Computer Graphics Forum, Vol. 25(4), pp. 789-807.

[15] Sorkine, O., Cohen-Or, D., Irony, D., Toledo, S. 2005. *Geometry-aware Bases for Shape Approximation*. IEEE, vol 11., No. 2.

[16] Toledo, S. 2003. TAUCS : A Library of Sparse Linear Solvers, version 2.2. Tel-Aviv University.

[17] Zhong, M., Qin, H., 2014. *Sparse Approximation of 3D Shapes via Spectral Graph Wavelets*. USA : Springer.

Optimal Strategy for Integrated Dynamic Inventory Control and Supplier Selection in Unknown Environment via Stochastic Dynamic Programming

Sutrisno[1], Widowati[2] and Solikhin[3]

[1,2,3]Department of Mathematics, Diponegoro University, Semarang, Indonesia

[1]tresno.math@undip.ac.id, [2]widowati_math@undip.ac.id, [3]soli_erf@yahoo.com

Abstract. In this paper, we propose a mathematical model in stochastic dynamic optimization form to determine the optimal strategy for an integrated single product inventory control problem and supplier selection problem where the demand and purchasing cost parameters are random. For each time period, by using the proposed model, we decide the optimal supplier and calculate the optimal product volume purchased from the optimal supplier so that the inventory level will be located at some point as close as possible to the reference point with minimal cost. We use stochastic dynamic programming to solve this problem and give several numerical experiments to evaluate the model. From the results, for each time period, the proposed model was generated the optimal supplier and the inventory level was tracked the reference point well.

Keywords: Inventory control problem, stochastic dynamic programming, supplier selection.

1. Introduction

Logistics and supply chain management involve optimal strategy determining to manage the procurement, movement and storage of materials, parts and finished product [1]. Inventory system and supplier selection are two of many components on logistics and supply chain management. Inventory level of a product has to be controlled for demand satisfying and the optimal supplier has to be selected if there are several alternatives. We can classify the environment of the inventory control problem into two class which are deterministic environment and stochastic environment.

Many researchers were developed the mathematical model to solve the inventory control problem in deterministic environment. For single product case, [2, 3, 4] were determined the optimal strategy for inventory controlling without supplier selection problem solving. In the other hand, [5, 6, 7] were solved dynamic supplier selection problem, but the inventory of the product was not controlled. Some researchers were integrated the inventory control problem and supplier selection problem where some assumptions were used. For example, [8] was solved the supplier selection problem integrated with inventory system but the inventory level was not controlled to follow the decision maker desiring, [9] was integrated production control with multi supplier and [10] was integrated the optimal control problem and supplier selection problem for single product under deterministic environment. Some researchers were also developed the mathematical model for multi-product inventory system, for example, [11] was solve the inventory control for it but the supplier selection problem was not solved and it was solved under deterministic environment.

Content from this work may be used under the terms of the Creative Commons Attribution 3.0 licence. Any further distribution of this work must maintain attribution to the author(s) and the title of the work, journal citation and DOI.
Published under licence by IOP Publishing Ltd

In this paper, we will formulate a mathematical model in the stochastic dynamic optimization form to determine the optimal strategy for an integrated single product inventory control problem and supplier selection problem in unknown environment, i.e. when the demand and purchasing cost parameters are random. We will use the stochastic dynamic programming to solve the proposed model to decide the optimal supplier from several alternatives and calculate the optimal product volume purchased from the optimal supplier so that the inventory level will be located at some point as close as possible to the reference point with minimal cost. The reference point will be decided by the decision maker. The Stochastic dynamic programming will be performed by LINGO optimization tool. We will give some numerical experiments to evaluate and simulate the model. From the results, for each time period, we will observe how the evolution of the inventory level compared to the reference point and who the optimal supplier is.

2. Another section of your paper

Let S denotes the number of the supplier. Let $X_{t,s}$ denotes the product volume purchased from supplier s at time period t. Let I_t denotes the inventory level of the product at time period t. The cost components in this problem are purchasing cost and holding cost. Let $U_{t,s}$ be the purchasing cost for the product per unit from supplier s at time period t and H_t be the holding cost per product's unit at time period t. Finally, let D_t denotes the random variable for the demand of the product.

To control the inventory level such that it will be as close as possible to a reference level r_t at time period t, we define the track reference objective $\|I_t - r_t\|_Q$ where Q is a quadratic semidefinite positive matrix and $\|x\|_Q$ means $x'Qx$ where x' is transpose of vector x. Let $Q=1$, then we define the cost function as follows

$$\sum_{i=1}^{\Omega}\left[\Pr_i \cdot \left(\sum_{t=1}^{T}\sum_{s=1}^{S}\sum_{p=1}^{P} U_{t,s}X_{t,s} + \sum_{t=1}^{T}\sum_{p=1}^{P} H_t I_t + \sum_{t=1}^{T}(I_t - r_t)^2\right)\right] \tag{1}$$

where $\Pr_i$ be the probability of scenario i. To ensure that the demand of each product is satisfied at each time period, then we have the first constraint as follows

$$I_{t-1} + \sum_{s=1}^{S} X_{t,s} - I_t \geq D_t, \forall t \in T. \tag{2}$$

Suppose that supplier s has maximum capacity C_s to supply product for every time period t, then we have the second constraint as follows

$$X_{t,s} \leq C_s, \forall t \in T, \forall s \in S. \tag{3}$$

For every time period, if the inventory has maximum capacity M, then we have the third constraint as follows

$$I_t \leq M, \forall t \in T. \tag{4}$$

Integer constraint for the purchased product volume can be written as follows

$$X_{t,s} \in \{0,1,2,\ldots\}, \forall t \in T, \forall s \in S. \tag{5}$$

Hence the mathematical model of this problem can be rewritten as follows

$$\min \sum_{i=1}^{\Omega} \left[\Pr_i \cdot \left(\sum_{t=1}^{T} \sum_{s=1}^{S} \sum_{p=1}^{P} U_{t,s} X_{t,s} + \sum_{t=1}^{T} \sum_{p=1}^{P} H_t I_t + \sum_{t=1}^{T} (I_t - r_t)^2 \right) \right] \tag{6}$$

subject to:

$$I_{t-1} + \sum_{s=1}^{S} X_{t,s} - I_t \geq D_t, \forall t \in T, \tag{7}$$

$$X_{t,s} \leq C_s, \forall t \in T, \forall s \in S, \tag{8}$$

$$I_t \leq M, \forall t \in T, \tag{9}$$

$$X_{t,s} \in \{0,1,2,\ldots\}, \forall t \in T, \forall s \in S. \tag{10}$$

3. Stochastic Dynamic Programming

Dynamic programming or multi-stage programming can be used to solve many optimization problems with several characteristics that are the problem can be divided into stages with the decision for each stage is required, each stage has several associate states, decision at a stage describe how the state at current stage is transformed into the state for the next stage, it has principle of optimality and there is recursion function for its stages [12]. Dynamic programming can be divided into two class which are deterministic, where all of information are known with certainty, and stochastic, where at least one parameter is uncertain/random. Stochastic dynamic programming can be solved by generating the scenarios of the problem and working forward/backward iteration.

To illustrate the stochastic dynamic programming, let t denote the stage usually time period of the problem, T denotes the number of the stage, x_t denotes the decision at time period t and Ω_t denotes the event space at time period t. The initial decision is taken in stage 0 and the recourse decisions is taken in stages 2, 3, …, T. A scenario is used to express the possible outcomes, usually cost/profit, based on the realization of the probability distribution. A scenario is presenting one possible realization of the future and the enumeration of all possible combinations of outcomes. For discrete event space, all of scenarios, called scenario tree, can be illustrated by Figure (1). For continuous or discrete but infinite event space, the scenario tree can be generated by using a sampling method, Monte Carlo sampling for example, to approximate the possible outcomes using a finite scenario tree [13].

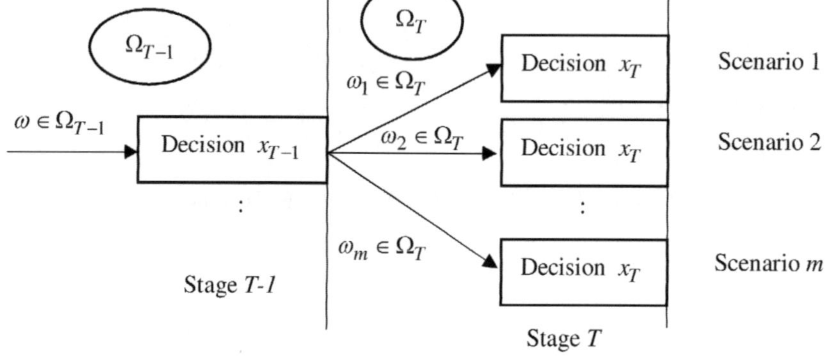

Figure 1. Scenario tree of a dynamic stochastic programming with discrete event space

Several optimization tools can be used to solve a dynamic programming by generating the scenario tree. In this paper, we use LINGO to solve the stochastic dynamic optimization (6).

4. Computational Simulation

4.1. The probability distribution is discrete

For the first numerical experiment, assume that the probability distribution of the random variables are discrete. Suppose that there are two suppliers namely s_1 and s_2. We will solve this problem for 3 periods. Suppose that the purchasing cost from supplier s_1 is \$12 per unit for each of time periods 1, 2 and 3. For time period-1, the purchasing cost from supplier s_2 is \$10 per unit, but for each of time periods 2 and 3, the purchasing cost is random with probability distribution given by Table (1).

Table 1. Probability distribution of purchase cost from s_2 for time period 2 and 3

Probability	Purchase cost (\$/unit/period)
0.5	10
0.5	12

The supplier capacity for each of suppliers s_1 and s_2 is 200 units per period. Let the initial inventory level is 0 item, the holding cost of the product is \$1/unit/period, the warehouse's maximum capacity is 200 units/period and the reference inventory level is 100 units for all time period. The demand for period 1 is known that is 100 units and the demand for each of periods 2 and 3 is random with the probability distribution given by Table (2).

Table 2. Demand's probability distribution for period 2 and 3

Probability	Demand (unit/period)
0.5	100
0.3	120
0.2	150

The decision maker desires that the inventory/stock level must be located at the reference level as close as possible with minimal cost. We solved the optimization problem (6) in Windows 8 AMD A6 2.7GHz and 2GB of Memory by using software LINGO 15.0 where the stochastic model class is multi-stage stochastic and the model class is MIQP (mixed integer quadratic programming). This stochastic dynamic program has 36 scenarios as shown by Table (3).

Table 3. Scenarios of the problem

Scenario	Time period (t)	Purchasing cost for s_2 (\$)	Demand (unit)	$X_{t,s}$		Inventory (unit)	Probability	Total Cost (\$)
				s_1	s_2			
1	1	12	100	0	199	99	0.0625	4270
	2	10	100	0	100	99		
	3	10	100	0	96	95		
2	1	12	100	0	199	99	0.0625	4490
	2	10	100	0	100	99		
	3	12	100	100	0	99		
3	1	12	100	0	199	99	0.0375	4470
	2	10	100	0	100	99		
	3	10	120	0	115	94		
4	1	12	100	0	199	99	0.0375	4700
	2	10	100	0	100	99		
	3	12	120	115	0	94		
				⋮				
36	1	12	100	0	199	99	0.01	5862
	2	12	150	149	0	98		
	3	12	150	146	0	94		

The initial solution i.e. optimal strategy for time period 1 is 0 unit purchased from s_1 and 199 units purchased from s_2. For time period 2, the optimal strategy can be decided after the event of the random variables in time period 2 is revealed. Finally, the optimal strategy for time period 3 can be decided after the event of the random variables in time period 3 is revealed. From the result, the total expected cost is \$4819 obtained by the probability and the total cost for all scenarios, that is

$$0.0625 \cdot 4270 + 0.0625 \cdot 4490 + 0.0375 \cdot 4470 + 0.0375 \cdot 4700 + \cdots + 0.0100 \cdot 5862 = 4819.$$

To illustrate how the optimal strategy for each time period is decided, let the demand for period 2 is 100 units and purchasing cost for s_2 for time period 2 is \$10/unit, then we have to purchase 0 unit from s_1 and 100 units from s_2 and inventory level for period 2 is 99 units. Let the demand for period 3 is 120 units and purchasing cost for s_2 is \$12/unit, then we have to purchase 115 units from s_1 and purchase 0 unit from s_2. This solution is illustrated by Table (4).

Table 4. Illustration of the solution

Time period (t)	Purchasing cost for s_2 (\$)	Demand (unit)	$X_{t,s}$		Inventory	Reference Inventory
			s_1	s_2		
1	12	100	0	199	99	100
2	10	100	0	100	99	100
3	12	120	115	0	94	100

From the comparison between inventory/stock level and the reference inventory level, it can be seen that the inventor/stock level of the product is followed the desired level well.

4.2. Demand's Probability Distribution is Continue

For the second example, suppose that the purchasing cost from s_2 for each of time periods 2 and 3 is normally distributed with mean 10 and standard deviation 4, the demand for each of periods 2 and 3 is normally distributed with mean 100 and standard deviation 20 and the remaining parameters are the same the first example. We solve this problem using LINGO where the scenarios are generated with Monte Carlo sampling where the sample size is 4. From the result, the total expected cost is \$4445.

For sensitivity analysis, suppose that the demand is normally distributed with mean 100 and standard deviation $\sigma > 0$. To observe the impact of the demand's standard deviation to the total expected cost, we evaluate the model with $\sigma = 5$, 10, 15, 20, 25, and 30. The total expected cost is given by Figure (2).

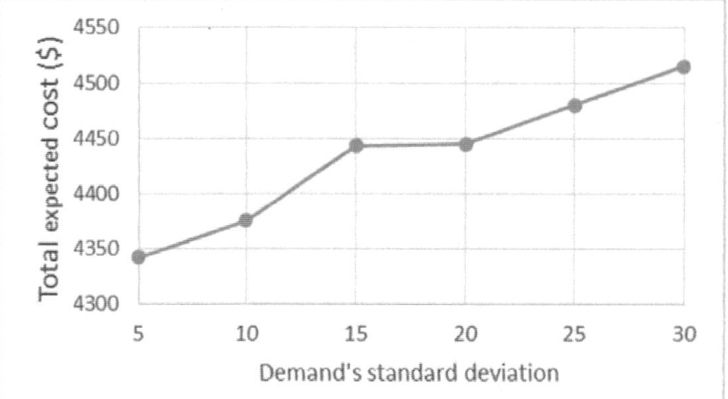

Figure 2. Impact of demand's standard deviation

From Figure (2), it can be seen that if the demand's standard deviation increases then the total expected cost is also increases. This is caused by the increasing of the demand uncertainty or the demand value has a wide range. If the demand's standard deviation more than 30, we predict that the total expected cost will be increased.

5. Concluding Remarks and Future Works

In this paper, the optimal strategy to control a single product dynamic inventory system integrated with supplier selection problem in unknown environment was considered. The mathematical model was written in the stochastic dynamic optimization and the optimal strategy was determined by using stochastic dynamic programming. Numerical experiments were considered with two cases which are the probability distribution of the random variables are discrete and continue in the normal probability distribution. From the results, it can be conclude that the supplier selection problem was solved and the inventory/stock level followed the desired level very well.

In the next works, we will develop the mathematical model for multi-product case along with order cost, penalty cost for defect or late delivery, service level, etc. Afterwards the random variable can be extended so that other parameters are random. In other works, we will develop the mathematical model so that it can also solve carrier selection problem.

References

[1] M. Christopher, Logistics & Supply Chain Management, Great Britain: Pearson Education Limited, 2011.

[2] Y. Arda and J.-C. Hennet, "Inventory control in a multi-supplier system," Int. J. Production Economics, vol. 104, pp. 249-259, 2006.

[3] W. Guerrero, T. Yeung and C. Guéret, "Joint-optimization of inventory policies on a multi-product multi-echelon pharmaceutical system with batching and ordering constraints," European Journal of Operational Research, vol. 231, pp. 98-108, 2013.

[4] S. Minner, "Multiple-supplier inventory models in supply chain management: A review," Int. J. Production Economics, Vols. 81-82, pp. 265-279, 2003.

[5] N. R. Ware, S. P. Singh and D. K. Banwet, "A mixed-integer non-linear program to model dynamic supplier Selection Problem," Expert Systems with Applications, vol. 41, pp. 671-678, 2014.

[6] C. Haksever and J. Moussourakis, "A model for optimizing multi-product inventory systems with multiple constraints," International Journal of Production Economics, vol. 97, pp. 18-30, 2005.

[7] S. S. Kara, "Supplier selection with an integrated methodology in unknown environment," Expert Systems with Applications, vol. 38, pp. 2133-2139, 2011.

[8] C. Guo and L. Xueping, "A multi-echelon inventory system with supplier selection and order allocation under stochastic demand," Int. J. Production Economics, vol. 151, pp. 37-47, 2014.

[9] A. Hajji, A. Gharbi, J.-P. Kenne and R. Pellerin, "Production control and replenishment strategy with multiple suppliers," European Journal of Operational Research, vol. 208, pp. 67-74, 2011.

[10] Sutrisno and P. A. Wicaksono, "Optimal Strategy for Multi-product Inventory System with Supplier Selection by Using Model Predictive Control," in Procedia Manufacturing of Industrial Engineering and Service Science (IESS), Yogyakarta, 2015.

[11] S. M. Mousavi, S. Niaki, A. Bahreininejad and S. N. Musa, "Multi-Item Multiperiodic Inventory Control Problem with Variable Demand and Discounts: A Particle Swarm Optimization Algorithm," The Scientific World Journal, vol. 2014, p. 16, 2014

[12] W. L. Winston, Operations Research: Applications and Algorithms, 1987.

[13] L. S. Team, LINGO: The Modeling Language and Optimizer, Illinois: Lindo Systems Inc., 2014.

Dissolution Behaviour of Hazardous Materials from Steel Slag with Wet Grinding Method

Nik Hisyamudin Muhd Nor[1], Siti Norhana Selamat[1], Muhammad Hanif Abd Rashid[1], Mohd Fauzi Ahmad[1], Saifulnizan Jamian[1], Sia Chee Kiong[1], Mohd Fahrul Hassan[1], Fariza Mohamad[1], Seiji Yokoyama[2]

[1]Department of Materials and Design Engineering, Faculty of Mechanical and Manufacturing Engineering, Universiti Tun Hussein Onn Malaysia, 86400, Parit Raja, Batu Pahat, Johor, Malaysia.
[2]Faculty of Mechanical Engineering, Toyohashi University of Technology, 1-1 Hibarigaoka Tempaku Toyohashi Aichi Japan.

E-mail: nhisyam@uthm.edu.my, sitinorhanaselamat@gmail.com, cd120111@siswa.uthm.edu.my, mohdfauzi@uthm.edu.my, saifulnz@uthm.edu.my, sia@uthm.edu.my, fahrul@uthm.edu.my, farizamd@uthm.edu.my, yokoyama@me.tut.ac.jp

Abstract. Steel slag is a by-product from steel industry and it contains variety of hazardous materials. In this study, the dissolution behaviour and removal potential of hazardous materials from steel slag with the wet grinding method was investigated. The slag was wet ground in the CO_2 atmosphere and the slurry produced was filtered using centrifugal separator to separate the liquid and solid sediments. Then, the concentrations of dissolved metal elements in the liquid sediment were analyzed by ICP-MS. The changes of pH during the grinding were also investigated. It was found that the pHs were decreased immediately after the CO_2 gas introduced into the vessel. The pHs were ranging from 6.8 to 7.6 at the end of grinding. The dissolved concentration of Zn and Cr were ranging from 5~45 [mg/dm^3] and 0.2~2.5 [mg/dm^3] respectively. The ratios of Zn removal for stainless steel oxidizing and reducing slag were very high, but those from normal steel oxidizing and reducing slag were very low. It is assumed that the Zn dissolved as $ZnOH^+$ from $Zn(OH)_2$ that formed due to the reaction between ZnO and water. Dissolution of Cr also occurred but in very low quantity compared to the dissolution of Zn. The dissolution of Cr occurred due to the grinding process and small amount of $Cr(OH)_3$ was formed during the grinding. This small formation of $Cr(OH)_3$ resulted to the low dissolved concentration of Cr in the form of $Cr(OH)_2^+$. According to the XRD analysis, the Cr mostly existed in the slags as Cr(III) in the form of $MgCr_2O_4$ and $FeCr_2O_4$.

1. Introduction

The steel industry is one of the key industries in the world due to the high demand especially from building construction, automotive and machinery sectors. Based on the World Steel Association (WSA) report, the world crude steel production in 2008 was approximately 1326.5 million tons [1]. Along with the production of steel, the steel industry also generates steelmaking slag as a by-product. These slags can be broadly divided into blast furnace slag and steel slag. In the blast furnace (BF), constituents other than iron in the iron ore are melted and become slag together with the ash of coke,

Content from this work may be used under the terms of the Creative Commons Attribution 3.0 licence. Any further distribution of this work must maintain attribution to the author(s) and the title of the work, journal citation and DOI.
Published under licence by IOP Publishing Ltd

and the slag was separated from the pig iron and recovered. This BF slag has components similar to those of natural rocks, and about 290 kg is generated per ton of pig iron. On the other hand, the steelmaking process which consists of refining pig iron, scrap and other material to produce steel leads to the production of steel slag. It can be divided into basic oxygen furnace (BOF) slag and electric arc furnace (EAF) slag with generation amounts of 110 and 120 kg per ton of crude steel respectively [2]. The EAF steel slag consisted of two different types of slag which are the oxidizing and reducing slag, with generation amounts of 70 and 50 kg per ton of crude steel respectively.

Steelmaking slag contains heavy metals such as chromium and zinc. The ingestion of these heavy metals beyond permissible quantities causes various chronic disorders in human beings [3-8]. The excessive ingestion of this substance into human body can affect several serious disorder such as genotoxicity, mutagenicity, and cell transformation, and lung cancer [3]. Alok Prasad Das et al. and Gibb H.J. et al. has reported that the risk of lung cancer among workers of the chromate industry in US increased compared to the general population and a positive correlation between duration of exposure and lung cancer death was found [4, 5].

Therefore, the removal of chromium and zinc from steelmaking slag has received a great deal of attention for recovery and elimination of those metals [9-12]. N.K. Hamadi et al. have investigated about the batch removal of hexavalent chromium (Cr(VI)) from wastewater under different experimental conditions using economic adsorbents [9]. The adsorbents were produced from the pyrolysis and activation of the waste tyres (TAC) and from the pyrolysis of sawdust (SPC). They found that the removal was favoured at low pH, with maximum removal at pH=2 for all types of carbon. All sorbents were found to efficiently remove Cr(VI) from solution.

In the previous studies, we have conducted several experiments especially in investigating the potential of slag as a sorbent for CO_2 and investigating the dissolution behavior of slag in pure water [13-15]. The wet grinding method was found to be applicable in dissolving and removing heavy metals inside the slag such as zinc and chromium. Therefore, our current research objective is to investigate the dissolution behavior and the removal potential of chromium and zinc with the wet grinding method.

2. Experimental

2.1. Samples

EAF oxidizing and reducing slag from normal and stainless steelmaking process were used in this experiment. The chemical compositions of the slag used are shown in Table 1. The “Oxi” and “Red” indicate oxidizing and reducing respectively. The Oxi C and Red B slag are the EAF normal steelmaking slag, while the others are the stainless steelmaking slag. Oxi A, Oxi B, Red A and Red B are rich with CaO and SiO_2, while Oxi C rich with FeO. The content of Cr_2O_3 in Oxi A, Oxi B, Oxi C and Red A were high but in Red B was very low. The content for Zn were low in all type of slags.

Table 1. Chemical compositions of the used slag [mass%].

Slag type	Size	Chemical composition [mass%]											
	[mm]	CaO	SiO_2	Al_2O_3	MgO	CaS	Cr_2O_3	MnO	TiO_2	FeO	NiO	CuO	ZnO
Oxi A	<10	35.4	36.3	13.5	6.9	0.2	2.1	3.3	2.0	0.23	0.03	0.06	0.01
Oxi B	<10	33.0	44.1	5.4	7.7	0.3	3.3	1	1.0	0.7	0.06	0.2	0.01
Oxi C	<10	20.5	20.2	20.1	6.2	0	1.7	4.1	0	26.8	0.03	0.03	0.07
Red A	<5	47.1	30.5	10.5	5.8	2.5	1.3	0.47	0.4	1.2	0.06	0.01	0.01
Red B	<5	50.3	21.1	18.7	4.1	4.0	0.2	0.6	0	0.9	0.02	0.07	0.01

According to the XRD analysis as in the previous works [13-16], Cr is mostly exists in the slag as Cr(III) in the form of $MgCr_2O_4$ and $FeCr_2O_4$. While, the compounds of Zn were not detected by XRD assumed to be due to its content which are too small. The example of cross section of the slag used is shown in Figure 1. Figure 1 shows the cross section of the EAF stainless steelmaking oxidizing slag (Oxi A). Whitish phases as in Figure 1(a); ①, ③, ⑤, and (b); ③, ⑤, are the phases that rich with Cr and Mg. The compositions of these elements are equivalent to that in $MgCr_2O_4$. The dark whitish phases as in Figure 1 (a)④ and (b)② are rich with the Cr and Fe, and the compositions of them are equivalent to that in $FeCr_2O_4$.

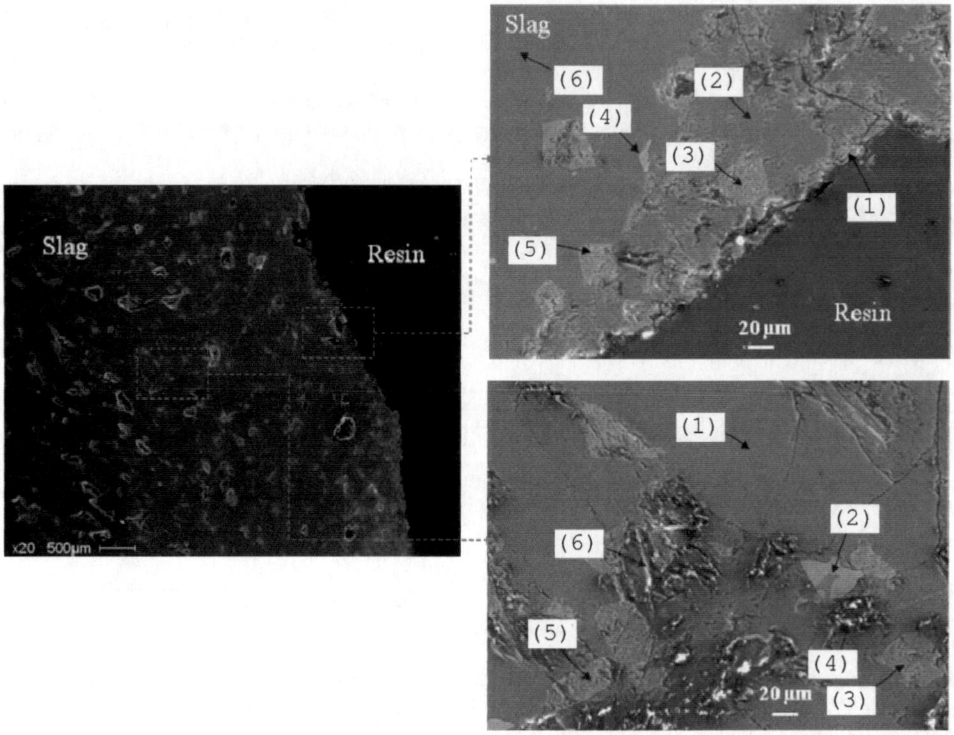

Figure 1. Cross section of EAF oxidizing slag (a) outer part, (b) inner part.

2.2. Procedure

The experimental apparatus used is as shown in Figure 2. Certain amount of slag and the alumina balls were placed in the grinding vessel. After the air in the vessel was evacuated to less than 1.3 Pa by a rotary pump, CO_2 gas was introduced into the vessel. This vacuum displacement was repeated twice to assure that all air in the vessel was fully replaced by CO_2. After that, pure water was poured into the vessel under a CO_2 gas flow. Finally, CO_2 gas was introduced until the internal pressure reached the desired value. Subsequently, the grinding was started.

The slag and the water became slurry under the grinding process. The pH of the slurry was measured with a pH meter. The slurry was then separated into the solid products and the aqueous solution using a centrifugal separator (HSIANGTAI Centrifuge, CN-820) at 1,500 rpm for 10 minutes. After the separated solid product was dried in air, it was characterized by SEM (JSM 6300, JEOL) and XRD (RINT 2500, Rigaku). The concentrations in the supernatant liquid were measured by ICP-AES

(P-4010, Hitachi). Finally, the liquid was vaporized, and the obtained sediments were dried in a laboratory dryer for one hour at 420 K. The sediments were then characterized by SEM and XRD.

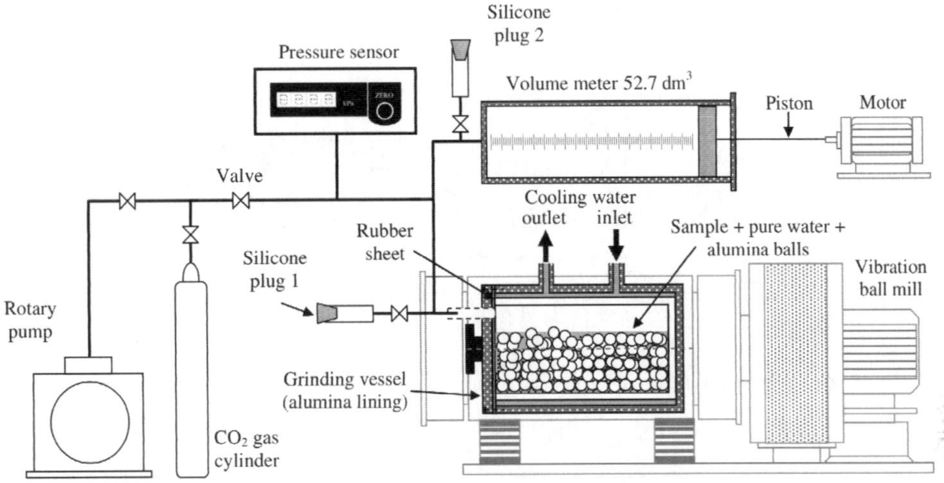

Figure 2. Schematic of the experimental apparatus.

The experimental conditions are shown in Table 2. The experiments were performed by changing the diameter of alumina ball, the initial pressure of CO_2, weight of the slag, and the volume of water. All of the grinding experiments were carried out at room temperature.

Table 2. Experimental conditions.

Condition	Value/Item
Sample	Oxi A, Oxi B, Oxi C, Red A, Red B
Diameter of alumina ball, d_b [mm] ( ) : number of ball	20 (180)
Partial pressure of CO_2, PCO_2 [kPa]	101.3
Weight of sample, W_s [kg]	0.2
Volume of water, V_w [dm^3]	0.4

3. 3. Results and Discussion

3.1. Change of pH with Grinding Time

Figure 3 shows the change in pH of the slurry for all samples during grinding in the CO_2 atmosphere. The initial pH indicates the pH of mixture of slag and water right after they had been mixed together. The initial pH ranging from 8 to 11.5 and it seemed to depend on the content of CaO in the slag. The higher the content, the higher the initial pH was. The pHs were decreased immediately after the grinding was started and almost unchanged in the latter stage of grinding for all samples. The pHs were ranging from 6.8 to 7.6 at the end of grinding. The decrease in the pH might occurred due to the dissolution of CO_2 gas into water, hydration of aqueous CO_2 and the ionization of hydrated CO_2, forming either carbonate or bicarbonate ions which can be summarized by the following equation:

$$CO_{2(g)} + H_2O \rightarrow HCO_3^- + H^+ \quad (1)$$

$$CO_{2(g)} + H_2O \rightarrow CO_3^{2-} + 2H^+ \quad (2)$$

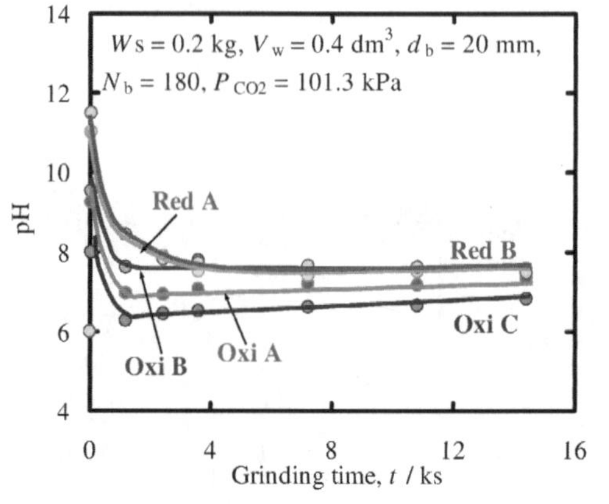

Figure 3. Change of pH during grinding process for all samples.

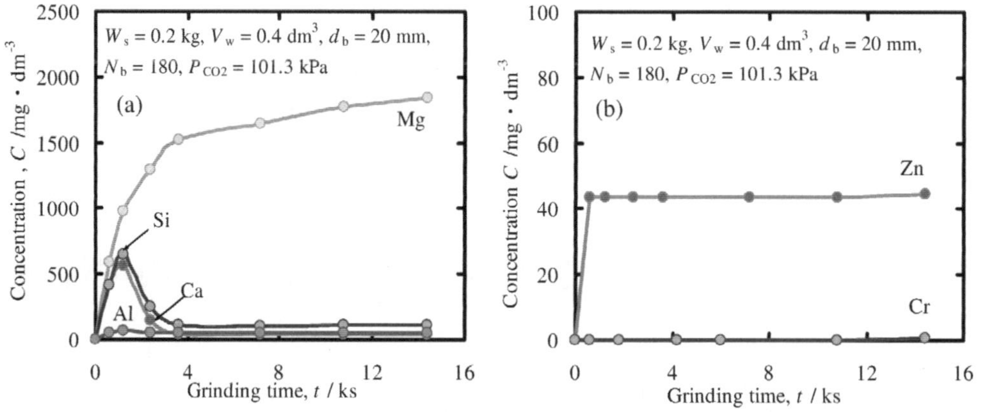

Figure 4. Dissolution behavior of (a) Ca, Mg, Si and Al, (b) Zn and Cr, during grinding. (Sample : Oxidizing A)

The change of pH during grinding affected the dissolution behavior of each element inside the slag. Figure 4 shows the example of dissolution behavior of elements inside the slag into water for oxidizing A. The dissolution of elements such as Ca, Mg, Si and Al were quite high. One of the Ca and Mg phases exist in the slag was $2CaO \cdot MgO \cdot 2SiO_{2(s)}$ [14, 15]. This phase might reacted with the following equation:

$$2CaO \cdot MgO \cdot 2SiO_{2(s)} + 5H_2O \rightarrow 2Ca^{2+} + Mg^{2+} + 2H_2SiO_{3(s)} + 6OH^- \quad (3)$$

Along with the dissolution of these elements, the dissolutions of Zn and Cr were also occurred during the grinding as in Figure 4 (b). The concentration of dissolved Zn was higher than that of Cr. The concentration of Zn for all samples is shown in Figure 5. The concentrations of Zn were instantly increased as the grinding started. However, they did not increase further and almost unchanged afterward except that for the reducing A. The concentration of Zn for reducing A increased for the second time after 10.8 ks before decreased back after that. In general, the dissolved concentrations of Zn from the oxidizing A, oxidizing B and reducing A which were the EAF stainless steelmaking slag were higher than those from the oxidizing C and reducing B slag which were the EAF steel slag from normal steelmaking process.

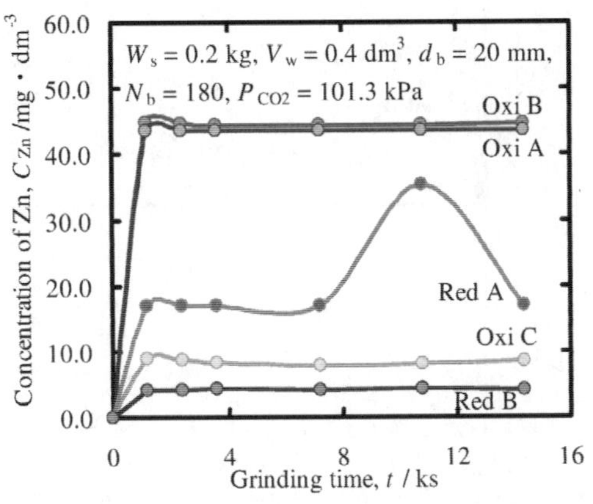

Figure 5. Dissolution behavior of Zn during grinding for all samples.

Using the composition of Zn inside each sample, m_{Zn} [mass%], weight of sample used, W_s [g], dissolved concentration of Zn, C_{Zn} [mg/dm^3] from Figure 5 and volume of water used, V_w [dm^3], the extraction rate of Zn, ER_{Zn} [%] can be defined as follows:

$$ER_{Zn} = \{[(C_{Zn} \times V_w) / 10^3] / (m_{Zn} \times W_s)\} \times 100 \quad (4)$$

Table 3 shows the removal ratio of Zn during grinding under CO_2 atmosphere. The ratios for Oxi A, Oxi B and Red A were very high, but those from Oxi C and Red B were very low. It can be concluded that Zn can be removed from slag using the mechanical grinding in CO_2 atmosphere. However, in the case of Oxi C and Red B, further investigation is needed to verify the reason for low removal ratio. $Zn(OH)_2$ that was assumed to be formed due to the reaction between ZnO and water dissolved as $ZnOH^+$ into the water as the following reaction:

$$ZnO_{(s)} + H_2O = Zn(OH)_{2(s)} \quad (5)$$
$$Zn(OH)_{2(s)} + H^+ = ZnOH^+ + H_2O \quad (6)$$

Table 3. Removal ratio of Zn [%].

Element	Oxi A	Oxi B	Oxi C	Red A	Red B
Zn	100.0	81.4	3.2	88.2	9.6

3.2. Dissolution behavior of Cr from slag

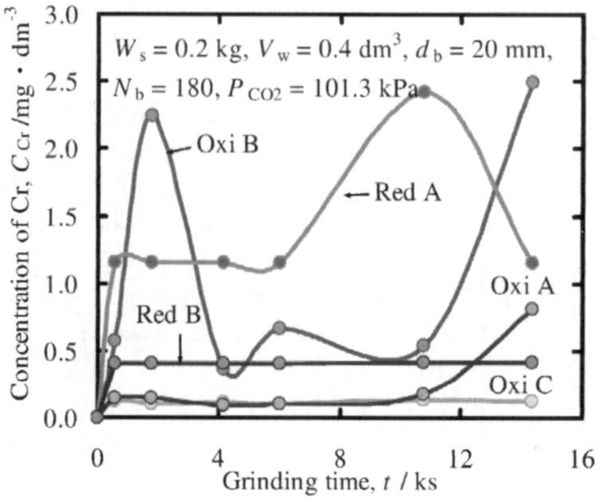

Figure 6. Dissolution behavior of Cr during grinding for all samples.

Figure 6 shows the dissolution behavior of Cr during grinding for each sample. The concentration of dissolved Cr for oxidizing A and reducing B increased slightly as the grinding started and almost unchanged afterward. The oxidizing A also increased slightly after the grinding was started but increased after 14.4 ks. The concentration of dissolved Cr from oxidizing B and reducing A were higher than the others. However, in comparison to the dissolved concentration of Zn shown in Figure 5, the concentrations of Cr from all samples were very small. Using the composition of Cr inside each sample, m_{Cr} [mass%], weight of sample used, W_s [g], dissolved concentration of Cr, C_{Cr} [mg/dm^3] from Figure 6 and volume of water used, V_w [dm^3], the extraction rate of Cr, ER_{Cr} [%] can be defined as follows:

$$ER_{Cr} = \{[(C_{Cr} \times V_w) / 10^3] / (m_{Cr} \times W_s)\} \times 100 \quad (7)$$

From this equation, the maximum extraction rates of Cr for oxidizing A, oxidizing B, oxidizing C, reducing A and reducing B were found to be 0.00023%, 0.00044%, 0.000044%, 0.0011% and 0.0014% respectively.

It is assumed from the concentration-pH diagram that the Cr dissolved into the aqueous solution as $Cr(OH)_2^+$ ion. It might be occurred according o the following reaction:

$$Cr_2O_3 + 3H_2O \rightarrow 2Cr(OH)_{3(s)} \quad (8)$$
$$Cr(OH)_{3(s)} + H^+ = Cr(OH)_2^+ + H_2O \quad (9)$$

This reaction occurred due to the grinding process and the amount of $Cr(OH)_3$ formed must be very small. This small formation of $Cr(OH)_3$ resulted to the low dissolved concentration of Cr in the form of $Cr(OH)_2^+$. According to the XRD analysis, the Cr mostly exists in the slags as Cr(III) in the form of $MgCr_2O_4$ and $FeCr_2O_4$.

4. Conclusion

From the results above, it can be concluded that:

1. The pHs were decreased immediately after the grinding was started and almost unchanged in the latter stage of grinding for all samples. The pHs were ranging from 6.8 to 7.6 at the end of grinding.

2. Extraction rate for Zn was high by this method. $Zn(OH)_2$ that was assumed to be formed due to the reaction between ZnO and water dissolved as $ZnOH^+$ into the water.
3. Dissolution of Cr occurred by the wet grinding method, but the dissolved concentrations were very low. It is assumed to be dissolved as $Cr(OH)_2^+$ ions from $Cr(OH)_3$ that was produced during the grinding.

5. Acknowledgements
This work is financially supported by the Research Acculturation Grant Scheme (RAGS) under the Ministry of Higher Education Malaysia.

6. References
[1] The Technical Society ISIJ 2010 Production and Technology of Iron and Steel in Japan during 2009, ISIJ International **50 (6)** 777–796.
[2] The Japan Iron and Steel Federation 2006 Nippon Slag Association The Slag Sector in the Steel Industry.
[3] IARC 1989 Monograph on Chromium, Nickel and Welding, Vol. **49**, International Agency for Research on Cancer, Lyon, France.
[4] Alok Prasad Das, Susmita Mishra 2008 Hexavalent Chromium (Vi) : Environment Pollutant and Health Hazard, Journal of Environmental Research And Development **2(3)** 386-392.
[5] Gibb H.J., Lees P.S. and Pinsky P.F., Rooney BC 2006 Lung cancer among workers in chromium chemical production, Am J Ind Med **38(2)** 115-126.
[6] Jamie L. Young, Sandra S. Wise, Hong Xie, Cairong Zhu, Tomokazu Fukuda, John Pierce Wise 2015 Comparative cytotoxicity and genotoxicity of soluble and particulate hexavalent chromium in human and hawksbill sea turtle (Eretmochelys imbricata) skin cells, Comparative Biochemistry and Physiology Part C: Toxicology & Pharmacology **178** 145-155.
[7] Nazma Shaheen, Nafis Md. Irfan, Ishrat Nourin Khan, Saiful Islam, Md. Saiful Islam, Md. Kawser Ahmed 2016 Presence of heavy metals in fruits and vegetables: Health risk implications in Bangladesh, Chemosphere **152** 431-438.
[8] Qihang Wu, Haichao Zhou, Nora F.Y. Tam, Yu Tian, Yang Tan, Song Zhou, Qing Li, Yongheng Chen, Jonathan Y.S. Leung 2016 Contamination, toxicity and speciation of heavy metals in an industrialized urban river: Implications for the dispersal of heavy metals, Marine Pollution Bulletin **104(1–2)** 153-161.
[9] N. K. Hamadi, X.D. Chen, M.M. Farid, M.G.Q. Lu 2001 Adsorption kinetics for the removal of chromium(VI) from aqueous solution by adsorbents derived from used tyres and sawdust, Chemical Engineering Journal **84 (2)** 95-105.
[10] S. Rengaraj, K.H.o Yeon, S.H. Moon 2001 Removal of chromium from water and wastewater by ion exchange resins, Journal of Hazardous Materials **87 (1–3)** 273-287.
[11] N.K. Lazaridis, D.D. Asouhidou 2003 Kinetics of sorptive removal of chromium(VI) from aqueous solutions by calcined Mg–Al–CO3 hydrotalcite, Water Research **37(12)** 2875-2882.
[12] V.K. Gupta, S. Agarwal, T. A. Saleh 2011 Chromium removal by combining the magnetic properties of iron oxide with adsorption properties of carbon nanotubes, Water Research **45(6)** 2207-2212.
[13] M.N.N. Hisyamudin, S. Yokoyama, M. Kawakami, M. Umemoto 2009 Reaction between CO2 and CaO under dry grinding, Powder Technology Elsevier **196 (2)** 156-162.
[14] M.N.N. Hisyamudin, S. Yokoyama, M. Umemoto 2010 Storage of CO2 in Low Al_2O_3 EAF Oxidizing Slag by Grinding with Vibration Mill, Materials Science Forum **654-656** 2927-2930.
[15] Yokoyama, S., Suzuki, A., Nor, N.H.B.M., Kanematsu, H., Ogawa, A., Takahashi, T., Izaki, M., Umemoto 2010 M., Serial batch elution of electric arc furnace oxidizing slag discharged from normal steelmaking process into fresh water, Tetsu-To-Hagane/Journal of the Iron and Steel Institute of Japan **96 (12)** 698-705.

Carbon Sequestration and Carbon Capture and Storage (CCS) in Southeast Asia

Nik Hisyamudin Muhd Nor[1], Siti Norhana Selamat[1], Muhammad Hanif Abd Rashid[1], Mohd Fauzi Ahmad[1], Saifulnizan Jamian[1], Sia Chee Kiong[1], Mohd Fahrul Hassan[1], Fariza Mohamad[1], Seiji Yokoyama[2]

[1]Department of Materials and Design Engineering, Faculty of Mechanical and Manufacturing Engineering, Universiti Tun Hussein Onn Malaysia, 86400, Parit Raja, Batu Pahat, Johor, Malaysia.
[2]Faculty of Mechanical Engineering, Toyohashi University of Technology, 1-1 Hibarigaoka Tempaku Toyohashi Aichi Japan.

E-mail: nhisyam@uthm.edu.my, sitinorhanaselamat@gmail.com, hd150060@siswa.uthm.edu.my, mohdfauzi@uthm.edu.my, saifulnz@uthm.edu.my, sia@uthm.edu.my, fahrul@uthm.edu.my, farizamd@uthm.edu.my, yokoyama@me.tut.ac.jp

Abstract. Southeast Asia is a standout amongst the most presented districts to unnatural weather change dangers even they are not principle worldwide carbon dioxide (CO_2) maker, its discharge will get to be significant if there is no move made. CO_2 wellsprings of Southeast Asia are mainly by fossil fuel through era of power and warmth generation, and also transportation part. The endeavors taken by these nations can be ordered into administrative and local level. This paper review the potential for carbon catch and capacity (CCS) as a part of the environmental change moderation system for the Malaysian power area utilizing an innovation appraisal structure. The country's recorded pattern of high dependence on fossil fuel for its power segment makes it a prime possibility for CCS reception. This issue leads to gradual increment of CO_2 emission. It is evident from this evaluation that CCS can possibly assume a vital part in Malaysia's environmental change moderation methodology gave that key criteria are fulfilled. With the reason to pick up considerations from all gatherings into the earnestness of an Earth-wide temperature boost issue in Southeast Asia, assume that more efficient measures can be taken to effectively accomplish CO_2 diminishment target.

1. Introduction

Long before the Industrial Revolution in the late of 18th century, human beings had only a moderate need for energy. They mostly relied on the energy and strength from human or animal to do work, and relied on the natural sources like wind and land to move something [1, 2]. To keep on living, human beings and all other living creatures require energy in the form of food and heat that were usually supplied by food, and sometimes by sunlight, wind, hot water and many more. Human also use energy for a number of other purposes, such as clothing, shelter, transportation, entertainment, cooling and the construction of tools [3].

Content from this work may be used under the terms of the Creative Commons Attribution 3.0 licence. Any further distribution of this work must maintain attribution to the author(s) and the title of the work, journal citation and DOI.
Published under licence by IOP Publishing Ltd

However, rapid growth in industrialization in the late 18th century has triggered the start of the Industrial Revolution in Europe and spread to other part of the world. Steam engines and other machinery required coal as fuel source and this has released the first emissions of carbon dioxide (CO2) and other gases which has then led to the increase in the concentration of the greenhouse gas (GHG) in the atmosphere, declining of air quality in many cities and induced the global warming issue in the world [4].

Global warming issue is at a serious level overall including Southeast Asia, which comprises of Brunei, Myanmar, Cambodia, Indonesia, Laos, Malaysia, Philippines, Singapore, Thailand, and Vietnam. The expansion in worldwide production of carbon dioxide (CO_2) is the major driver of the global warming which is an issue of incredible concern today [1]. Anthropogenic climate change due to combustion of fossil fuel has been found as the primary cause of CO_2 discharges into the air that lead to global warming [2]. Keeping in mind the end goal to diminish the outflows of CO_2 numerous innovation alternatives have been proposed and investigated so as to accomplish a reasonable low carbon vitality society [3]. There is no single innovation choice accessible that could totally tackle the discharges issue of world vitality needs and in all probability an arrangement of innovation arrangements would be required to avoid perilous anthropogenic impedance with the atmosphere framework [4]. Southeast Asia region is extremely exposed against the environmental changes as a large proportion of the general population and a great part of the monetary action are located along the coastlines. Nations in the area are dynamic members in global environmental change arrangements and some have set national targets, Indonesia means to lessen GHG emanations by 26% from nothing new (BAU) levels by 2020, or 41% in the event that it gets upgraded worldwide help; Malaysia looks to diminish the carbon force of total national output (GDP) by 40% by 2020 from 2005 levels, and Singapore intends to decrease GHG outflows by 16% from BAU levels by 2020 as appeared in Table 1.

Table 1. Summary of energy policies in Southeast Asia countries [5]

Country	Sector	Policies and Targets
Brunei Darussalam	Efficiency	45% reduction in energy intensity compared to 2005 levels by 2035
	Renewables	electricity generation from renewable energies achieve 10% by 2035
Cambodia	Efficiency	20% reduction in energy consumption by 2035
Indonesia	Efficiency	1% reduction in energy intensity per year until 2025
	New and renewable energy	23% and 31% increase in sharing of "new and renewable energy" in primary energy by 2025 and 2050 respectively
	Climate change	26% reduction of GHG emission by 2020, and 41% reduction with the assistance from international communities
Laos PDR	Efficiency	10% reduction of final energy consumption
	Renewables	Renewable energies achieve 30% share by 2025
Malaysia	Efficiency	Promote improvement of the energy efficiency in industry, buildings and domestic sectors
	Renewables	Boost capacity of renewable energies to 2,080 MW by 2020 and 4,000 MW by 2030. Plans for renewable energy and, site selection and regulatory studies are developed.
	Nuclear	Plans for nuclear energy are developed and, site selection and regulatory studies are being carried out.
	Climate change	40% reduction in carbon intensity of GDP by 2020 (from 2005 levels).
Myanmar	Efficiency	10 % reduction in energy demand (from BAU level).
	Renewables	Renewable energy in total generation capacity achieves 15% to 18% share by 2020.

Philippines	Efficiency	Achieve energy savings equivalent to 15% of annual final demand relative to BAU by 2020
	Renewables	Triple the capacity of renewable power generation to 15GW by 2030.
Singapore	Efficiency	35% reduction in energy intensity by 2030 from 2005 levels
	Climate change	7% to 11% reduction in GHG emissions by 2020. The rate will be increased to 16%, if there is a legal binding agreement globally on climate change. 36% reduction in GHG emissions intensity by 2030 from 2005 levels.
Thailand	Efficiency	30% reduction in energy intensity by 2036 through the elimination of fossil-fuel subsidies and improvements on the energy efficiency.
	Renewables	20% increase in the power generation from renewable energy and 20% increase in the biofuels based transport by 2036.
	Nuclear	Two commercial nuclear reactors have been planned since 2007, although progress has delayed since the Fukushima Daiichi accident.
Vietnam	Efficiency	5 to 8 % reduction in energy consumption by 2015 and 8-10% by 2020.
	Renewables	4.5% and 6% increase in the share of renewables in electricity generation by 2020 and by 2030 resepectively.
	Nuclear	10.7 GW capacity of nuclear power is being developed until 2030.

Table 2 demonstrates the progressions of CO_2 outflow in term of tons per capita of these nations contrasted with Southeast Asia nations showing that Brunei, Singapore, Indonesia, Malaysia, and Thailand have comparable CO_2 emission with developed nations. The radical addition of the CO_2 outflow was identified with the flourishing mechanical advancement in the nations, for example, Indonesia, Malaysia, Thailand and Vietnam. Also, there is a misguided judgment that CO_2 discharge from Southeast Asia is less significant contrasted with the real emitters, United States and China. Indonesia, with the fourth most elevated populace on the planet, contributed 4.73% of aggregate world nursery gasses discharge[6]. It is an interest of quick regard for the CO_2 outflow in this area. The greatest test and weight confronted by Southeast Asia nations now is to find a determination to keep the advancement and an Earth-wide temperature boost relief in parity [5], [7]. In order to achieve mid to long term CO_2 reduction targets, cost effective CO_2 capture from fossil fuel power plant and subsequent sequestration options need to be evaluated keeping in view of world's growing demand of energy. There are several technologies available for carbon capture and storage (CCS) such as absorption, adsorption, gas separation, membrane and cryogenic separation. The aim of this literature review is to provide a current understanding of the potential of various CCS technologies for Southeast Asia.

Table 2. CO_2 emission in Southeast Asia [5].

Country	CO_2 Emission (tons per capita)		Variation (%)
	1990	2009	
Cambodia	0.005	0.33	560.0
Brunei	25.47	23.69	-7.0
Indonesia	0.81	1.90	134.6
Laos	0.06	0.30	400.0
Malaysia	3.11	7.10	128.3
Myanmar	0.11	0.23	109.1
Philippines	0.68	0.75	10.3
Singapore	15.41	6.39	-58.5
Thailand	1.68	3.95	135.1
Vietnam	0.32	1.65	415.6

2. CCS development in Southeast Asia

2.1. CCS Background
Essentially, there are three principle segments of the CCS process called catch, transport and capacity. The catch step includes catching, getting dried out, and compacting CO_2 from vast stationary discharge sources. For fuel-blazing procedures, for example, people with significant influence plants, partition advances can be utilized to catch CO_2 after ignition or to decarbonize the fuel before burning. The vehicle step requires through tankers, pipeline or boat to exchange caught CO_2 to a suitable stockpiling site which typically situated at far from the CO_2 source. This is conceivable when the caught CO_2 gas is compacted to a high thickness at the catch office. Potential stockpiling strategies can be makeshift (infusion into underground topographical arrangements, infusion into the profound sea) or for all time (modern fixation in inorganic carbonates) [8]. Rubin *et al.* reports that just a few CCS were viewed as experienced advances, for example, pipeline, industrial separation and enhance oil recuperation. Additionally, sea storage still in research stage while mineral carbonation and geological storage considered at exhibit stage entering monetarily feasible.

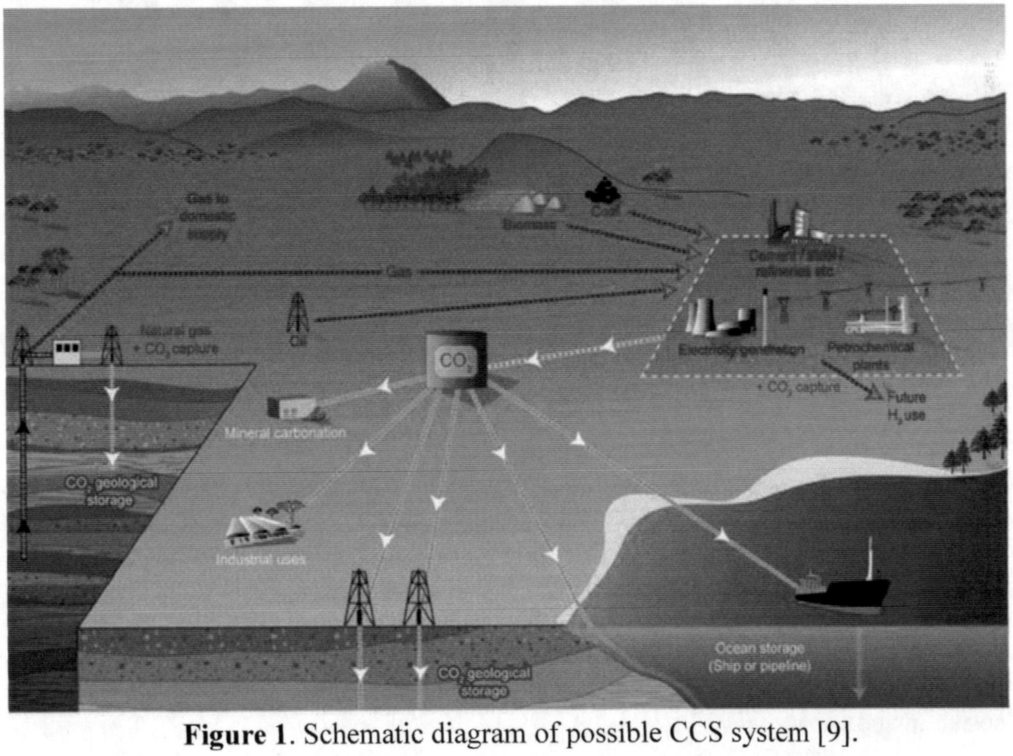

Figure 1. Schematic diagram of possible CCS system [9].

Malaysia second-biggest oil maker in Southeast Asia as of now records for more than 1/6 of Southeast Asia's aggregate vitality related CO_2 outflows, an offer that remaining parts stable until 2040. Its per-capita CO_2 discharges are more three times higher than the Southeast Asian normal over the Outlook period and more than one-and-a-half times higher than the OECD normal by 2040, which has been declining subsequent to 2007 and is relied upon to proceed [5], [6], [8]. However, Malaysia right now has restricted abilities in investigating and creating practical CO_2 partition innovations. There have been a set number of surveys on the potential and specialized possibility for CCS in Malaysia [10], [11]. There has been no expansive scale financed research activity for CCS in Malaysia and it is likely that the catch innovation as with most environmental change alleviation advancements would have been produced by industrialized countries [12].

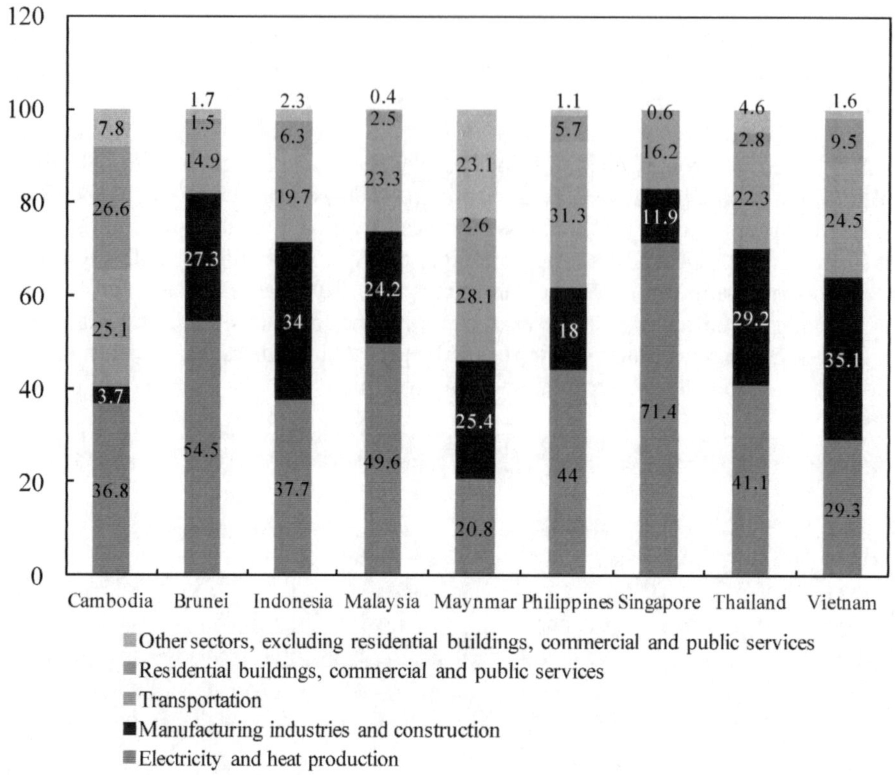

Figure 2. Energy-related CO2 emissions by sector in Malaysia [6].

2.2. CO_2 Capture Technology

Catching CO_2 can be accomplished utilizing one of the three principle strategies in particular by means of post-burning catch, pre-ignition catch or oxyfuel burning [12]. CO_2 in pipe gas can be isolated and caught from various stationary emanation sources, for example, fossil fuel power plants, bond creation, refineries, normal gas preparing, steel plants, and biomass plants. In some of these sources, for example, common gas–processing plants where CO_2 should be isolated from crude gas for creating deals quality regular gas, CO_2 partition is as of now a standard part of the procedure. Caught CO_2 needs just to be dried and packed before transportation and capacity.

2.3. CO_2 Storage Technology

There are a few potential storage choices, for example, the captured CO_2 infused into profound geographical developments, (for example, develop or exhausted oil and gas fields), profound coal creases that can't be mined, and saline aquifers. Infused CO_2 will be put away as a thick stage supercritical liquid and could be caught through various diverse components, for example, basic and stratigraphic catching, lingering gas catching, dissolvability catching, mineral catching, and hydrodynamic catching. A few issues of capacity would decide the most doable innovation to be grow, for example, the expense of capacity, including the expense of transportation from the source to the capacity site, ought to be minimized. The danger of mischances such gas spillage ought to be disposed of and ecological effect ought to be negligible. The capacity technique ought not abuse any national or worldwide laws and regulations. Yet, mineral carbonation is the main known type of changeless CO_2 stockpiling that is by all accounts essentially practical. Vast amounts of normal minerals and modern squanders are accessible for use in altering CO_2. Around 10–15% of the aggregate climatic CO_2 is transmitted from little and medium CO_2 emitters (<2.5 M ton CO_2), which are potential wellsprings of CO_2 for mineral carbonation [13], [14]. The thought of mineral carbonation was initially proposed by Seiftriz in 1990. Mineral carbonation, the response of metal bearing oxides with CO_2 to shape insoluble strong carbonates, emulates the regular weathering of silicate rocks, which happens over topographical time scales [15]. Besides, mineral carbonation on industrial by product such as steelmaking slag [13], [16]–[18], cement waste [19], [20] and fly ash should also be taken into consideration as the CO_2 emission mostly sourced from the industrial plant itself.

2.4. Costing for CCS System

For the plants with topographical stockpiling and no EOR credit, the expense of CCS reaches from 0.02–0.05 US$/kWh for PC plants and 0.01–0.03 US$/kWh for NGCC plants (both utilizing post-burning catch). For IGCC plants (utilizing pre-ignition catch), the CCS cost ranges from 0.01–0.03 US$/kWh in respect to a comparative plant without CCS. For all power frameworks, the expense of CCS can be diminished by around 0.01–0.02 US$/kWh when utilizing EOR with CO_2 stockpiling in light of the fact that the EOR incomes somewhat make up for the CCS costs. The biggest cost diminishments are seen for coal based plants, which catch the biggest measures of CO_2. In a few cases, the low end of the CCS cost extent can be negative, showing that the expected credit for EOR over the life of the plant is more noteworthy than the most reduced reported expense of CO_2 catch for that framework. This may likewise apply in a couple cases of ease catch from modern procedures. Table 3 indicates the summary of the expenses of CO_2 capture, transport and storage reported in altogether with monitoring expenses.

Table 3. Cost ranges for the components of a CCS system as applied to a given type of power plant or industrial source year 2002 [9].

CCS system components	Cost range	Remarks
Capture from a coal- or gas-fired power plant	15-75 US$/t$CO_2$ net captured	Net costs of captured CO_2, compared to the same plant without capture.
Capture from hydrogen and ammonia production or gas processing	5-55 US$/t$CO_2$ net captured	Applies to high-purity sources requiring simple drying and compression
Capture from other industrial sources	25-115 US$/t$CO_2$ net captured	Range reflects use of a number of different technologies and fuels.
Transportation	1-8 US$/t$CO_2$ transported	Per 250 km pipeline or shipping for mass flow rates of 5 (high end) to 40 (low end) MtCO_2 yr-1.
Geological storage	0.5-8 US$/t$CO_2$ net injected	Excluding potential revenues from EOR or ECBM
Geological storage: monitoring and verification	0.1-0.3 US$/t$CO_2$ injected	This covers pre-injection, injection, and post-injection monitoring, and depends on the regulatory

		requirements
Ocean storage	5-30 US$/t$CO_2$ net injected	Including offshore transportation of 100-500 km, excluding monitoring and verification
Mineral carbonation	50-100 US$/t$CO_2$ net mineralized	Range for the best case studied. Includes additional energy use for carbonation

ASEAN countries were considered as a rapid developing region where the energy demand projected to keep rising even though there are also efforts to reduce the nation energy consumption as mention in their policies. Malaysia through the nation Economic Transformation Plan (ETP). Through its strides towards modernization and industrialization, the nation has persistently appreciated a steady, solid power supply at generally low costs. The danger of environmental change would bring upon change towards this situation. The backing and cooperation towards GHG discharge decreases would be required from all countries. Malaysia is quick turning into an industrialized country and thusly, has a more noteworthy part in the battle against environmental change. CCS in its present structure is costly and later on in the wake of considering all the conceivable cost change would at present be more expensive than the current customary fossil fuel based vitality. This is a hard truth which is hard to be acknowledged by numerous. The lawmaker, the businesses and the overall population must be set up to adjust to an all the more unreasonable vitality situation if there is to be any genuine advancement towards environmental change alleviation.

Clearly, the battles to moderate environmental change including CCS may look extravagant. Nevertheless, with the start of environmental change, alleviation endeavors may avert future harm, which could cost considerably more [21]. Proper endeavors host been done by the gatherings included under the name of an Earth-wide temperature boost moderation. Be that as it may, accord is constantly difficult to be come to in this issue. From the workshops, gatherings, or meetings held, we can scarcely see a conceivable result from them. This condition is going on in Southeast Asia as well as the entire world. Case in point, Copenhagen Accord, which was considered as the continuation of Kyoto Protocol, turned out as a feebly bound collaboration, which its suggestion is misty.

3. Conclusion

This paper has generally described and reported efforts taken on global warming mitigation in Southeast Asia countries using CCS technology. In addition, environmental strategies achieve very little unless it is tied to socio-economic and trade or international obligations. These factors have made the efforts unable to achieve their expected result. Technology and fund transfer from developed countries is vital as an aid to these vulnerable countries while criticism will make no influence. Aside from that, in order to create a sustainable development along with global warming mitigation, more frequent internal cooperation among ASEAN countries is needed.

4. Acknowledgements

This work is financially supported by the Research Acculturation Grant Scheme (RAGS) under the Ministry of Higher Education Malaysia.

5. References

[1] M.W. Flinn, D. Stoker, 1700-1830: the Industrial Revolution, Clarendon Press 2, 1984.
[2] T.M. Geraghty, The factory system in the British industrial revolution: A complementarity thesis, European Economic Review 51(6) (2007) p.1329-1350.
[3] R.C. Neville, Energy Needs-Energy Sources, Solar Energy Conversion (Second Edition), 1995, p. 1-38.
[4] W. Bartok, A.F. Sarofim, Fossil Fuel Combustion-A Source Book, A Wiley-Interscience Publication, 1991.
[5] K. Kaku, "Global warming and climate change of Asian countries including Japanese domestic greenhouse gas (GHG) reduction in the filed of poultry and swine industries," Procedia Eng., vol. 8, pp. 511–514, 2011.
[6] M. K. Mondal, H. K. Balsora, and P. Varshney, "Progress and trends in CO 2 capture/separation technologies: A review," Energy, vol. 46, no. 1, pp. 431–441, 2012.
[7] A. Fazeli, F. Bakhtvar, L. Jahanshaloo, N. A. Che Sidik, and A. E. Bayat, "Malaysia no dnats s' municipal solid waste conversion to energy: A review," Renew. Sustain. Energy Rev., vol. 58, pp. 1007–1016, 2016.
[8] D. Y. C. Leung, G. Caramanna, and M. M. Maroto-Valer, "An overview of current status of carbon dioxide capture and storage technologies," Renew. Sustain. Energy Rev., vol. 39, pp. 426–443, 2014.
[9] I. E. Agency, "Southeast Asia Energy Outlook," World Energy Outlook Spec. Rep., p. 131, 2013.
[10] Z. H. Lee, S. Sethupathi, K. T. Lee, S. Bhatia, and A. R. Mohamed, "An overview on global warming in Southeast Asia: CO2 emission status, efforts done, and barriers," Renew. Sustain. Energy Rev., vol. 28, pp. 71–81, 2013.
[11] N. Y. G. Lai and E. H. Yap, "Carbon capture and storage for developing economies: Yhe case for Malaysia," 2011 IEEE Conf. Clean Energy Technol., pp. 182–186, 2011.
[12] ADB, Prospects for carbon capture and storage in southeast Asia. Asian Development Bankz, 2013.
[13] E. Rubin, L. Meyer, and H. de Coninck, "IPCC Special Report on Carbon Dioxide Capture and Storage: Technical Summary," Rite.or.Jp, 2005.
[14] M. R. Othman, Martunus, R. Zakaria, and W. J. N. Fernando, "Strategic planning on carbon capture from coal fired plants in Malaysia and Indonesia: A review," Energy Policy, vol. 37, no. 5, pp. 1718–1735, 2009.
[15] T. H. Oh, "Carbon capture and storage potential in coal-fired plant in Malaysia - A review," Renew. Sustain. Energy Rev., vol. 14, no. 9, pp. 2697–2709, 2010.
[16] N. Y. G. Lai, E. H. Yap, and C. W. Lee, "Viability of CCS: A broad-based assessment for Malaysia," Renew. Sustain. Energy Rev., vol. 15, no. 8, pp. 3608–3616, 2011.
[17] Y. Sun, M. S. Yao, J. P. Zhang, and G. Yang, "Indirect CO2 mineral sequestration by steelmaking slag with NH4Cl as leaching solution," Chem. Eng. J., vol. 173, no. 2, pp. 437–445, 2011.
[18] A. A. Olajire, "A review of mineral carbonation technology in sequestration of CO2," J. Pet. Sci. Eng., vol. 109, pp. 364–392, 2013.
[19] A. Azdarpour, M. Asadullah, R. Junin, M. Manan, H. Hamidi, and A. R. M. Daud, "Carbon dioxide mineral carbonation through ph-swing process: A review," Energy Procedia, vol. 61, pp. 2783–2786, 2014.
[20] M. N. Nik Hisyamudin, S. Yokoyama, and U. Minoru, "Storage of CO2 in Low Al2O3 EAF Oxidizing Slag by Grinding with Vibration Mill," Mater. Sci. Forum, vol. 654–656, pp. 2927–2930, 2010.
[21] K. Zhang, J. Liu, W. Liu, and J. Yang, "Preparation of glass-ceramics from molten steel slag using liquid-liquid mixing method," Chemosphere, vol. 85, no. 4, pp. 689–692, 2011.
[22] M. Barati, S. Esfahani, and T. A. Utigard, "Energy recovery from high temperature slags," Energy, vol. 36, no. 9, pp. 5440–5449, 2011.

[23] M. F. Rojas and M. I. Sánchez De Rojas, "Chemical assessment of the electric arc furnace slag as construction material: Expansive compounds," Cem. Concr. Res., vol. 34, no. 10, pp. 1881–1888, 2004.

[24] A. R. Brough and A. Atkinson, "Sodium silicate-based, alkali-activated slag mortars - Part I. Strength, hydration and microstructure," Cem. Concr. Res., vol. 32, no. 6, pp. 865–879, 2002.

[25] R. Ismail, "An economic evaluation of carbon emission and carbon sequestration for the forestry sector in Malaysia," Biomass and Bioenergy, vol. 8, no. 5, pp. 281–292, 1995.

Journal of Physics: Conference Series **725** (2016) 012011 doi:10.1088/1742-6596/725/1/012011

Optimal Control for TB disease with vaccination assuming endogeneous reactivation and exogeneous reinfection

N Anggriani[1], B C Wicaksono, A K Supriatna
Department of Mathematics
Faculty of Mathematics and Natural Sciences, Universitas Padjadjaran
Jl. Raya Bandung-Sumedang km 21 Jatinangor 45363, Indonesia

[1]E-mail: nursanti.anggriani@unpad.ac.id and nsanti_a@yahoo.co.id

Abstract. Tuberculosis (TB) is one of the deadliest infectious disease in the world which caused by *Mycobacterium tuberculosis*. The disease is spread through the air via the droplets from the infectious persons when they are coughing. The World Health Organization (WHO) has paid a special attention to the TB by providing some solution, for example by providing BCG vaccine that prevent an infected person from becoming an active infectious TB. In this paper we develop a mathematical model of the spread of the TB which assumes endogeneous reactivation and exogeneous reinfection factors. We also assume that some of the susceptible population are vaccinated. Furthermore we investigate the optimal vaccination level for the disease.

1. Introduction

Tuberculosis (TB) is among the fatal diseases known globally. It is estimated 9.6 millions new TB cases reported in 2014 [1]. Meanwhile, Indonesia is the third rank in regards to this new TB prevalence compared to other countries globally. In fact, TB vaccine is already available in the market. Since 1906 BCG (Bacille Calmette-Guerin) vaccine has been available. However the occurence of TB is still increasing from time to time. This is alarming, hence many effort have been explored to attack the TB transmission from various view of points, including from theoretical view point via the utilization of mathematical modeling to understand and to control the TB. Recently some mathematical models of TB transmission have been developed [2,3]. In this paper, based on earlier model [4] we derived a pulmonary and non-pulmonary transmission of TB as in [5]. Here we develop the mathematical model of the spread of the TB which assumes endogeneous reactivation and exogeneous reinfection factors. Here we also assume that some of the susceptible population are vaccinated, since vaccination is still common method in TB elimination [6]. These complexities are considred simultaneously in the model to give a better insight into the analysis of the optimal control of the TB transmission.

2. Model Formulation

Here we assume that human population is divided into seven compartments as the following: the number of susceptible compartment individual in time t is denoted by $S(t)$, the number of latent compartment individual in time t is denoted by $L(t)$, the number of vaccinated compartment individual

Content from this work may be used under the terms of the Creative Commons Attribution 3.0 licence. Any further distribution of this work must maintain attribution to the author(s) and the title of the work, journal citation and DOI.
Published under licence by IOP Publishing Ltd

in time t is denoted by V(t), the number of infectious/pulmonary *TB* compartment individual in time t is denoted by $I_i(t)$, the number of non-infectious/extra pulmonary *TB* compartment individual in time t is denoted by $I_n(t)$, the number of recovered from infectious *TB* compartment individual in the time t is denoted by $R_i(t)$ and the number of recovered from non-infectious *TB* compartment individual in the time t is denoted by $R_i(t)$. The model is represented by the diagrams shown in Figure 1, with the description of the parameters given in Table 1. The model assumes that the recruitment rate is coming from the birth (there is no immigration factor), individual who are vaccinated goes to L if they are not immune. Each compartment will experience natural death but to compartment infectious and non-infectious an extra factor of death caused by TB is added. The causes of TB disease are also assumed to be affected by endogeneous reactivation or exogeneous reinfection.

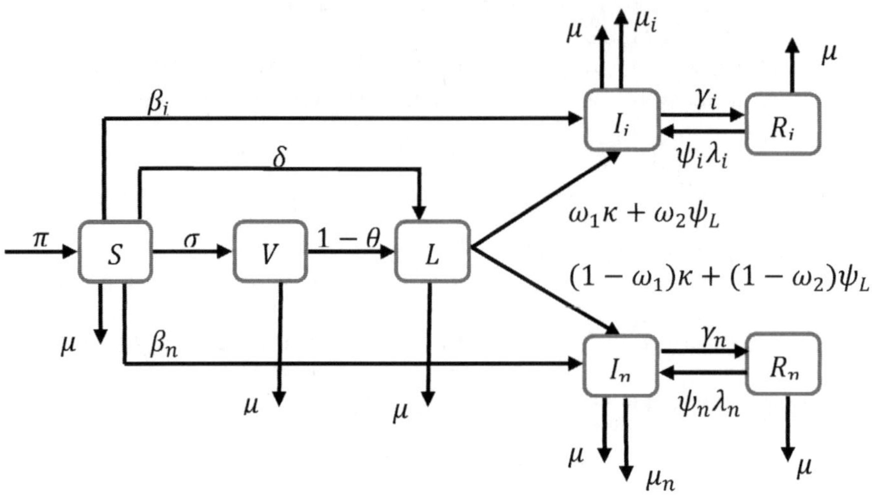

Figure 1. Schematic diagram of the TB disease transmission with endogeneous reactivation I_i and exogeneous reinfection I_n.

Based on the schematic diagram in Figure 1, we obtain the mathematical model of the TB disease transmission as follows:

$$\frac{dS}{dt} \quad \pi - \delta S I_i - \beta_i S I_i - \beta_n S I_i - \sigma S - \mu S \tag{1}$$

$$\frac{dV}{dt} \quad \sigma S - (1-\theta) V I_i - \mu V \tag{2}$$

$$\frac{dL}{dt} \quad \delta S I_i + (1-\theta) V I_i - \kappa L - \psi_L L I_i - \mu L \tag{3}$$

$$\frac{dI_i}{dt} \quad \beta_i S I_i + (\omega_1 \kappa + \omega_2 \psi_L I_i) L + \lambda_i \psi_i R_i I_i - \gamma_i I_i - \mu_i I_i - \mu I_i \tag{4}$$

$$\frac{dI_n}{dt} \quad \beta_n S I_i + ((1-\omega_1)\kappa + (1-\omega_2)\psi_L I_i) L + \lambda_n \psi_n R_n I_i - \gamma_n I_n - \mu_n I_n - \mu I_n \tag{5}$$

$$\frac{dR_i}{dt} \quad \gamma_i I_i - \lambda_i \psi_i R_i I_i - \mu R_i \tag{6}$$

$$\frac{dR_n}{dt} \quad \gamma_n I_n - \lambda_n \psi_n R_n I_i - \mu R_n \tag{7}$$

Table 1. Definition of parameters used in model

Parameter	Description
π	Birth rate
μ	Natural mortality rate
μ_i	Mortality rate of the infectious TB
μ_n	Mortality rate of the non-infectious TB
β_i	Transmission rate from susceptible compartment to infectious compartment I_i
β_n	Transmission rate from susceptible compartment to non-infectious compartment I_n
δ	Transmission rate from susceptible compartment to exposed compartment
σ	Vaccination rate for the susceptible
θ	Vaccination effectiveness
ψ_L, ψ_i, ψ_n	Rate of exogenous reinfection L, I_i, and I_n
ω_1	Fraction of exposed infectious TB
ω_2	Fraction of exposed non-infectious TB
κ	Rate of endogenous reactivation
γ_i	Recovery rate from infectious TB, I_i
γ_n	Recovery rate from non-infectious TB, I_n
λ_i	Transmission rate from recovered (R_i) compartment to I_i compartment
λ_n	Transmission rate from recovered (R_n) compartment to I_n compartment

2.1 *Equilibrium and the Reproduction Number*

From the model (1-7), we have two equilibrium points, i.e nonendemic,

$$(S, V, L, I_i, I_n, R_i, R_n) \quad \left(\frac{\pi}{(\sigma+\mu)}, \frac{\pi\sigma}{\mu(\sigma+\mu)}, 0,0,0,0,0\right)$$

and endemic point

$$S^* \quad \frac{\pi}{\beta_n I_i+\beta_i I_i+\delta I_i+\mu+\sigma},$$

$$V^* \quad -\frac{\sigma\pi}{a^*-b^*} \quad \frac{\sigma\pi}{(b^*-a^*)},$$

$$L^* \quad \frac{I_i\pi(\sigma\theta+\delta I_i\theta-\delta\mu-\delta I_i-\sigma)}{(a^*-b^*)(k+\psi_l I_i+\mu)},$$

$$R_i^* \quad \frac{\gamma_i I_i}{\lambda_i\psi_i I_i+\mu},$$

$$R_n^* \quad \frac{\gamma_n I_n}{\lambda_n\psi_n I_i+\mu},$$

with

$$a^* \quad \delta I_i^2\theta + \beta_i I_i^2\theta + \beta_n I_i^2\theta + \sigma I_i\theta + \mu I_i\theta$$

$$b^* \quad \delta I_i^2 + \delta I_i\mu + \beta_i I_i^2 + \beta_i I_i\mu + \beta_n I_i^2 + \beta_n I_i\mu + \sigma I_i + \sigma\mu + \mu I_i + \mu^2$$

The Reproduction Number for model (1 - 7) is given by :

$$\mathcal{R}_0 \quad \left(\frac{\beta_i\pi}{\sigma+\mu} + \frac{\omega_1 k(\mu\delta\pi+(1-\theta)\pi\sigma)}{\mu(\sigma+\mu)(k+\mu)}\right)\left(\frac{1}{\gamma_i+\mu_i+\mu}\right).$$

3. Model Analysis

In this section we analyze the model by assuming that there is a program of vaccination to the susceptibe population. We define the following objective function:

$$J(u) \quad \int_{t_0}^{t_1}(A_1 I_i(t) + A_2 I_n(t) + C_1 u_1^2(t) + C_2 u_2^2(t))dt$$

where t_1 is the final time and the coefficients A_1, A_2 are the weight factors for infectious population, u_1, u_2 are the rate of the vaccination and the effectiveness control with C_1 and C_2 are the balancing cost factor due to scale the importance of I. The constraints of the optimization problem for the Eqs. (1-7) are :

$t_0 < t < t_1,\ 0 \le u_1(t) \le 1,\ 0 \le u_2(t) \le 1,\ S(0) \quad S_0 \ge 0,\ V(0) \quad V_0 \ge 0,\ L(0) \quad L_0 \ge 0,$
$I_i(0) \quad I_{i0} \ge 0, I_n(0) \quad I_{n0} \ge 0, R_i(0) \quad R_{i0} \ge 0,$

Here we assume that there is a sufficient fund such that we are able to undertake a full vaccination program by vaccinating all susceptible in the human population, $u_1 = 1$. Hence, the Hamiltonian of the problem is given by

$$H \quad f(t,x,u) + \lambda g(t,x,u)$$
$$H \quad A_1 I_i(t) + A_2 I_n(t) + C_1 u_1^2(t) + C_2 u_2^2(t) + \lambda_1(\pi - \delta S I_i - \beta_i S I_i - \beta_n S I_i - u_1 S - \mu S) + \lambda_2(u_1 S - (1-u_2)V I_i - \mu V) + \lambda_3(\delta S I_i + (1-u_2)V I_i - \kappa L - \psi_L L I_i - \mu L) + \lambda_4(\beta_i S I_i + (\omega_1\kappa + \omega_2\psi_L I_i)L + \lambda_i\psi_i R_i I_i - \gamma_i I_i - \mu_i I_i - \mu I_i) + \lambda_5\big(\beta_n S I_i + \big((1-\omega_1)\kappa + (1-\omega_2)\psi_L I_i\big)L + \lambda_n\psi_n R_n I_i - \gamma_n I_n - \mu_n I_n - \mu I_n\big) + \lambda_6(\gamma_i I_i - \lambda_i\psi_i R_i I_i - \mu R_i) + \lambda_7(\gamma_n I_n - \lambda_n\psi_n R_n I_i - \mu R_n)$$

with $\lambda_1, \lambda_2, \lambda_3, \lambda_4, \lambda_5, \lambda_6,$ and λ_7 *co-state* variable or the Lagrange multiplier of the optimization problem. The necessary conditions that an optimal control must satisfy come from the Pontryagin's Maximum Principle. To accomplish this, we define the Hamiltonian H for the control problem as above, and define the state vector variables:

$$\dot{x}(t) \quad \begin{bmatrix} \dot{S}(t) \\ \dot{V}(t) \\ \dot{L}(t) \\ \dot{I}_i(t) \\ \dot{I}_n(t) \\ \dot{R}_i(t) \\ \dot{R}_n(t) \end{bmatrix}, \text{ and the vector variables for the co state equation : } \dot{\lambda}(t) \quad \begin{bmatrix} \dot{\lambda}_1(t) \\ \dot{\lambda}_2(t) \\ \dot{\lambda}_3(t) \\ \dot{\lambda}_4(t) \\ \dot{\lambda}_5(t) \\ \dot{\lambda}_6(t) \\ \dot{\lambda}_7(t) \end{bmatrix}.$$

Then the necessary conditions are:

$$\dot{S} \quad \frac{\partial H}{\partial \dot{\lambda}_1} \quad \pi - \delta S I_i - \beta_i S I_i - \beta_n S I_i - u_1 S - \mu S$$
$$\dot{V} \quad \frac{\partial H}{\partial \dot{\lambda}_2} \quad u_1 S - (1-u_2)V I_i - \mu V$$
$$\dot{L} \quad \frac{\partial H}{\partial \dot{\lambda}_3} \quad \delta S I_i + (1-u_2)V I_i - \kappa L - \psi_L L I_i - \mu L$$
$$\dot{I}_i \quad \frac{\partial H}{\partial \dot{\lambda}_4} \quad \beta_i S I_i + (\omega_1\kappa + \omega_2\psi_L I_i)L + \lambda_i\psi_i R_i I_i - \gamma_i I_i - \mu_i I_i - \mu I_i$$
$$\dot{I}_n \quad \frac{\partial H}{\partial \lambda_5} \quad \beta_n S I_i + \big((1-\omega_1)\kappa + (1-\omega_2)\psi_L I_i\big)L + \lambda_n\psi_n R_n I_i - \gamma_n I_n - \mu_n I_n - \mu I_n$$
$$\dot{R}_i \quad \frac{\partial H}{\partial \dot{\lambda}_6} \quad \gamma_i I_i - \lambda_i\psi_i R_i I_i - \mu R_i$$
$$\dot{R}_n \quad \frac{\partial H}{\partial \dot{\lambda}_7} \quad \gamma_n I_n - \lambda_n\psi_n R_n I_i - \mu R_n$$

with co-state equation :

$$\dot{\lambda_1} = -\frac{\partial H}{\partial S} = \lambda_1(\delta I_i + \beta_i I_i + \beta_n I_i + u_1 + \mu) - \lambda_2 u_1 - \lambda_3 \delta I_i - \lambda_4 \beta_i I_i - \lambda_5 \beta_n I_i$$
$$\dot{\lambda_2} = -\frac{\partial H}{\partial V} = \lambda_2\big((1-u_2)I_i + \mu\big) - \lambda_3((1-u_2)I_i)$$
$$\dot{\lambda_3} = -\frac{\partial H}{\partial L} = \lambda_3(\kappa + \psi_L I_i + \mu) - \lambda_4(\omega_1\kappa + \omega_2\psi_L I_i) - \lambda_5\big((1-\omega_1)\kappa + (1-\omega_2)\psi_L I_i\big)$$
$$\dot{\lambda_4} = -\frac{\partial H}{\partial I_i} = -A_1 + \lambda_1(\delta S + \beta_i S + \beta_n S) + \lambda_2\big((1-u_2)V\big) - \lambda_3(\delta S + (1-u_2)V - \psi_L L) - \lambda_4(\beta_i S + \omega_2\psi_L L + \lambda_i\psi_i R_i - \gamma_i - \mu_i - \mu) - \lambda_5(\beta_n S + (1-\omega_2)\psi_L L + \lambda_n\psi_n R_n) - \lambda_6(\gamma_i - \lambda_i\psi_i R_i) + \lambda_7(\lambda_n\psi_n R_n)$$
$$\dot{\lambda_5} = -\frac{\partial H}{\partial I_n} = -A_2 + \lambda_5(\gamma_n + \mu_n + \mu) - \lambda_7\gamma_n$$
$$\dot{\lambda_6} = -\frac{\partial H}{\partial R_i} = -\lambda_4(\lambda_i\psi_i I_i) + \lambda_6(\lambda_i\psi_i I_i + \mu)$$
$$\dot{\lambda_7} = -\frac{\partial H}{\partial R_n} = -\lambda_5(\lambda_n\psi_n I_i) + \lambda_7(\lambda_n\psi_n I_i + \mu)$$

The optimal condition can be achieved via the Pontryagin Maximum Principle, i.e., $\frac{\partial H}{\partial u_1} = 0$, which gives :

$$2C_1u_1 - \lambda_1 S + \lambda_2 S = 0\,,$$
$$2C_1u_1 = \lambda_1 S - \lambda_2 S\,,$$
$$u_1 = \frac{\lambda_1 S - \lambda_2 S}{2C_1} = \frac{(\lambda_1 - \lambda_2)S}{2C_1}\,,$$

with the conditions $0 \le u_1 \le 1$. Then finally we have:

$$u_1 = max\left\{min\left(\frac{(\lambda_1-\lambda_2)S}{2C_1}, 1\right), 0\right\}.$$

Similarly, $\frac{\partial H}{\partial u_2} = 0$ which gives:

$$2C_2u_2 + \lambda_2(VI_i) - \lambda_3(VI_i) = 0$$
$$2C_2u_2 = \lambda_3(VI_i) - \lambda_2(VI_i)$$
$$u_2 = \frac{\lambda_3(VI_i) - \lambda_2(VI_i)}{2C_2} = \frac{(\lambda_3-\lambda_2)VI_i}{2C_2}$$

with the conditions $0 \le u_2 \le 1$. Then finally we have:

$$u_2 = max\left\{min\left(\frac{(\lambda_3-\lambda_2)VI_i}{2C_2}, 1\right), 0\right\}.$$

Here the sufficient conditions for both control giving the minimum objective value are $\frac{\partial^2 H}{\partial u_1^2} = 2C_1 \ge 0$ and $\frac{\partial^2 H}{\partial u_2^2} = 2C_2 \ge 0$. In the next section we provide a numerical example of the result above to show that the controls are really exist.

4. Numerical Simulation

Numerical simulation is done to illustrate the theory above. Here we choose the parameters values described in Table 2 to run the simulation in the model. The results are presented in Figure 2(a) to 2(h).

Journal of Physics: Conference Series **725** (2016) 012011 doi:10.1088/1742-6596/725/1/012011

Table 2. The values of parameters used in the subsequent simulation of the model.

Parameter	Value
μ	0.2509710694
μ_i	0.6587000003
μ_n	0.5000000000
λ_i	0.7100995912
λ_n	0.5000000000
γ_i	0.2200000000
γ_n	0.2100000000
ω_1	0.1600000000
ω_2	0.7800000000
β_i	0.0027308696
β_n	0.0050000000
ψ	0.8500000000
δ	0.02883022718
κ	0.2522000000

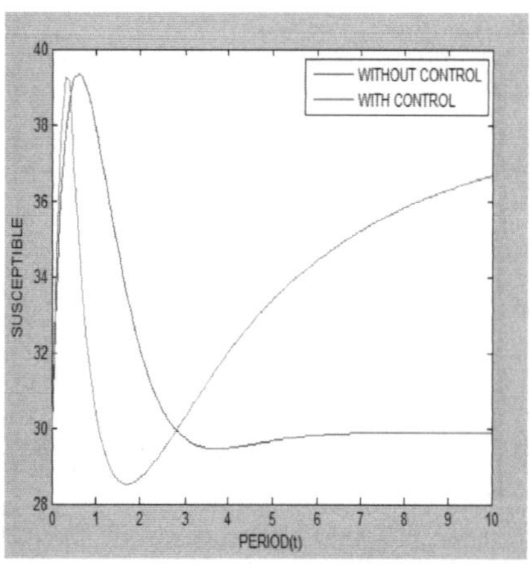

(a)

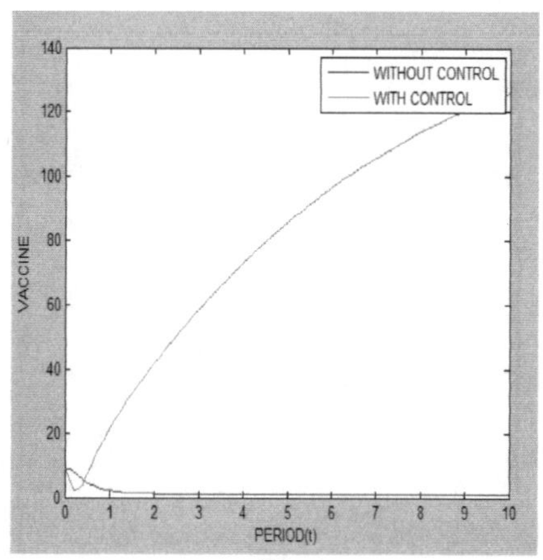

(b)

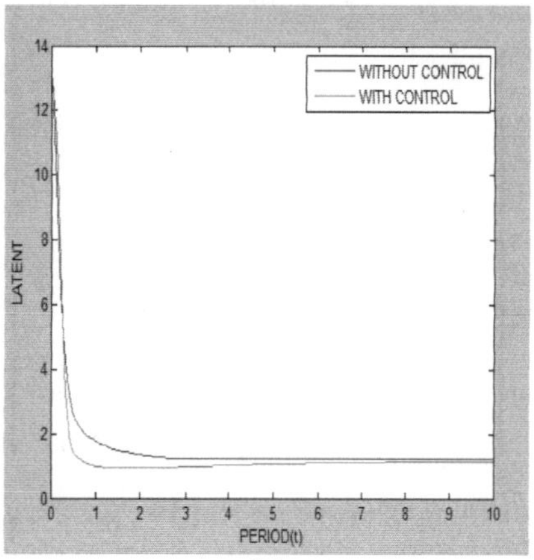

(c)

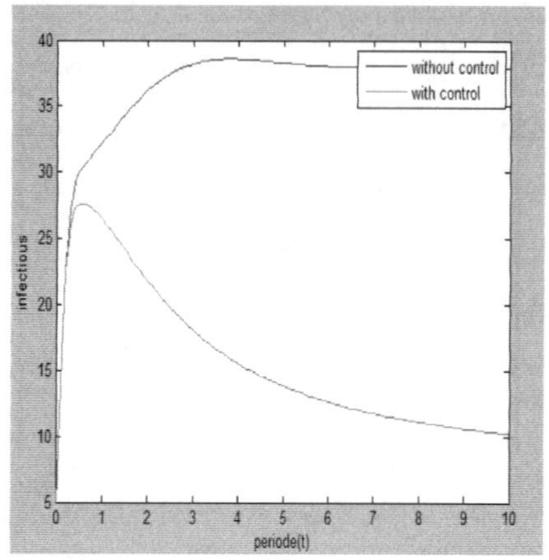

(d)

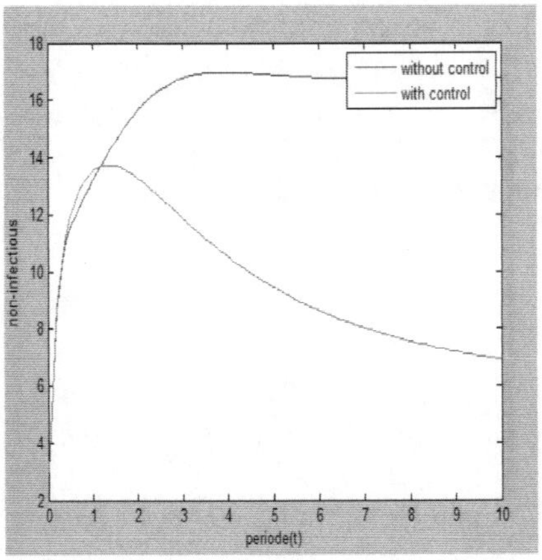

(e)

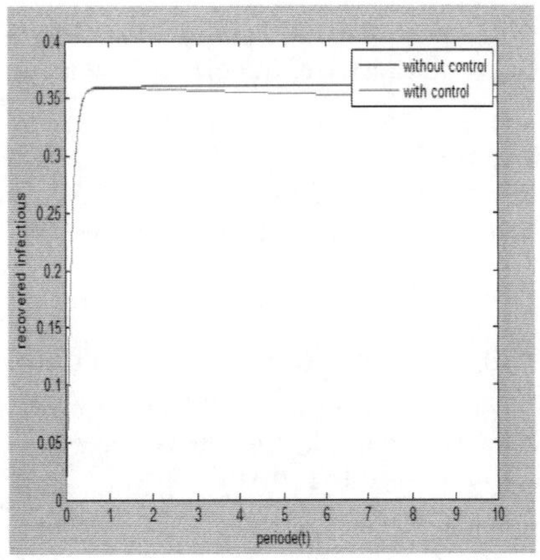

(f)

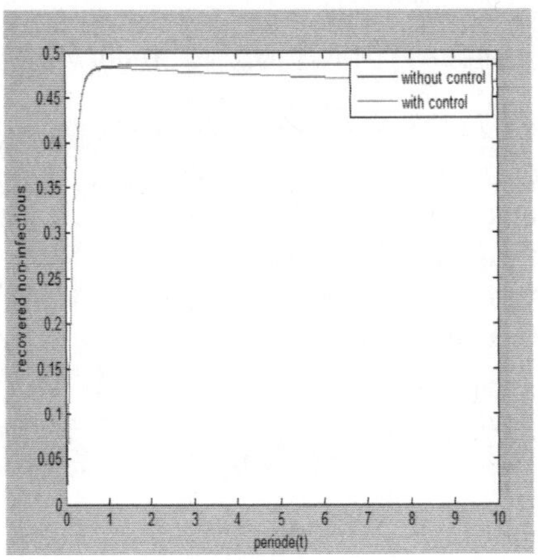

(g)

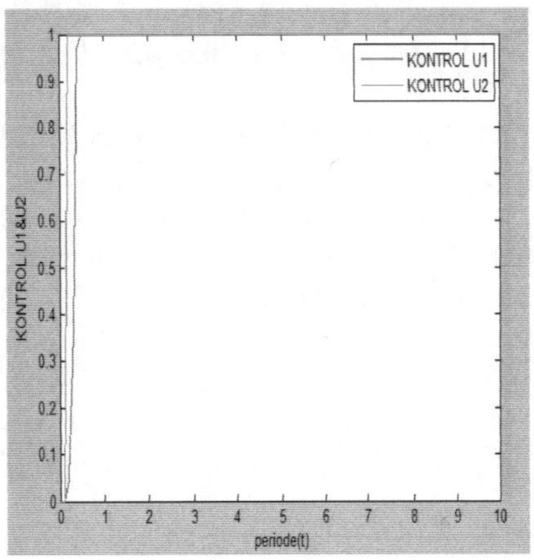

(h)

Figure 2. The results of the simulation for the data set given in Table 1.

Figure 2(a) shows the trajectory for the susceptible class. Figure 2(b) shows the trajectory for the vaccinated class. Figure 2(c) shows the trajectory for the latent. Figure 2(d) shows the trajectory for the infectious TB class. Figure 2(e) shows the trajectory for the non-infectious TB Class. Figure 2(f) shows the trajectory for the recovered from th infectious TB class and figure 2(g) for the non-infectious TB class. The control profiles are given in figure 2(h). The resulting Ro upon the implementation of the controls is 0.48. In the simulation we use the values: $A_1 = 1$, $A_2 = 1$, $C_1 = 1$, $C_2 = 1$. The results presented in the figure above indicate that when there is no vaccination, the basic reproduction number (R_0) is certainly greater than one. However, when we apply vaccination control program with the control profiles given by figure 2(h) then the basic reproduction number reduces to just 0.48. This shows the success of the control program as indicated by figure 2(d) and 2(c) for the decrease of the population in the infected compartment.

5. Conclusion

In this paper we have developed a mathematical model for the transmission of tuberculosis disease with the effect of control, i.e. by taking into account the factor of BCG vaccine, and considering the endogeneous reactivation and exogeneous reinfection. The optimal control to the TB disease by BCG vaccine aims to reduce the number of population of active TB by taking into account the lowest possible cost. Optimal control dynamic models of disease control tuberculosis vaccine in this paper concluded that vaccination is better done optimally to the susceptible population if the vaccine is highly effective. Otherwise, the effect of vaccination will not be optimal.

Acknowledgment
This work is funded by BOPTN Universitas Padjadjaran Indonesia under the Scheme of Penelitian Unggulan Perguruan Tinggi (PUPT) 2016

References

[1] WHO 2015 Global *Tuberculosis* Edition 20
[2] Nainggolan J, Supian S, Supriatna A K and Anggriani N 2013 *Mathematical Model of Tuberculosis Transmission with Reccurent Infection and Vaccination.* Journal of Physics: Conference Series 423 (2013) 012059.
[3] Nainggolan J, Supian S, Supriatna AK, Anggriani N and Detiatrimargini 2014 *Optimal Control Solution of A Tuberculosis Transmission Model with Reccurent Infection and Vaccination Using C# Programming* Advanced Science Letter American Scientific Publishers Vol. 20 p 51-55.
[4] Mishra, Bimal K and Srivastava J 2014 *Mathematical Model on Pulmonary and Multidrug-resistant Tuberculosis Patients with Vaccination* Journal of the Egyptian Mathematical Society 22 (2): 311-316.
[5] Anggriani N, Fithriati I, Agustian D, Supriatna AK 2015 *The Effect of DOTS Strategy on the Transmission of Tuberculosis with Endogenous Reactivation and Exogenous Reinfection Factors* Proceeding 3rd International Conference on Chemical, Agricultural and Medical Science (CAMS-2015) Dec 10-11, 2015, Singapura, http://dx.doi.org/10.15242/IICBE.C1215007.
[6] Egbetade S A and Ibrahim M O 2014 *Modelling the Impact of BCG Vaccines on Tuberculosis Epidemics* Journal of Mathematical Modelling and Application 9 (1):49-55.

AUTHOR INDEX